Jairo José Rondón Contreras

Petroleum tar pitch

Jairo José Rondón Contreras

Petroleum tar pitch

An experimental study as a binder

ScienciaScripts

Imprint

Cover image: www.ingimage.com

This book is a translation from the original published under ISBN 978-613-9-40755-2.

Publisher:
Sciencia Scripts
is a trademark of
Dodo Books Indian Ocean Ltd. and OmniScriptum S.R.L publishing group

120 High Road, East Finchley, London, N2 9ED, United Kingdom
Str. Armeneasca 28/1, office 1, Chisinau MD-2012, Republic of Moldova, Europe
Managing Directors: Ieva Konstantinova, Victoria Ursu
info@omniscriptum.com

Printed at: see last page
ISBN: 978-620-8-54838-4

Contents

DEDICATION

To my wife Alicia and my daughter Paula, my two loves.

ACKNOWLEDGEMENTS

To my parents, a source of admiration, to whom I owe thanks for giving me life and forming me as an independent man, instilling

for giving me life and shaping me into an independent man, instilling in me the conduct and moral

moral firmness that directs my path through life. I know they will always be proud of me. proud of me.

To my wife Alicia and my daughter Paula for filling my life with love and joy.

To my colleagues who always supported me in the realisation of this book.

To Dr. Héctor Del Castillo, who was very attentive to this project,

and his support and knowledge at all times, thank you very much.

To the Illustrious University of Los Andes for giving me the opportunity to

to develop myself in an integral way.

PROLOGUE

In this book, you will find information on the production of petroleum tar pitch (PTP) from vacuum gas oil (VGO) for use as a binder for carbon anodes in the aluminium industry, an experimental matrix for its production and a thermodynamic model to evaluate its conversion from vacuum gas oil to petroleum tar pitch.

On the first pages you will find a synthesis of everything related to thermal cracking from a refining point of view and the concepts of vacuum gas oil and oil tar pitch, written in such a way that interested people who are just starting out in the world of refining are immediately familiarised with the technical processes.

You will then see an experimental methodology for obtaining petroleum tar pitch base under moderate severity conditions. You will also get a thermodynamic model based on a C-C bond breaking mechanism, with free radical formation, including also bridge-type breaks between aromatic polycondensed systems, as in the case of resins and asphaltenes.

Last but not least you will find an analysis of the properties of vacuum gas oil and petroleum tar pitch, together with conclusions on pitch base production, following this brief methodology.

On the other hand, you will know that the main idea of the writer of this book is to share his experiences so that, through them, you as a reader can easily understand the fundamental concepts and properties necessary to obtain petroleum tar pitch. We hope that you will enjoy this book and that its application will be used or serve as a seed for new ideas in the oil and aluminium industry.

Yours sincerely,

Héctor L. Del Castillo P., Ph.D.
University of Los Andes
Venezuela

INTRODUCTION

The world oil industry is currently developing research into refining processes aimed at increasing the production of high commercial value streams from heavy crude oil and the residues generated as a result of refinery processing. Typically, the products obtained in a heavy crude oil refinery are: Liquefied Petroleum Gas (LPG), Gasoline, Diesel, Kerosene, Jet Fuel, Petroleum Coke, Fuel Oil and other diluents; streams that are mostly marketed as energy resources that are necessary for the daily operation and industrial development on a global scale. Although the refining business has been oriented almost exclusively to the production of fuels, other by-products and intermediate streams can be used in the petrochemical industry, in the aluminium industry and even for the production of advanced carbon materials. Formulating and implementing technological alternatives for the treatment of refinery waste streams arises from the need to revalue these by-products in the global market, in order to generate greater income opportunities for the oil industry. This is the case of heavy or vacuum gas oils (GOV), produced by thermal cracking processes, whose low quality means that in many cases it is used as a diluent and/or industrial fuel.

Some of these gas oils have aromatic fractions rich in condensed rings, therefore their potential application has been considered to produce petroleum tar pitch (BAP; which serves as a binder for carbon anodes used in aluminium production), using thermal cracking processes, which involve free radical production, polymerisation reactions, condensation, dehydrogenation and dealkylation, without using catalysts for this purpose.

This book is presented following a procedure to identify and simulate the physical and chemical conditions for their optimal formation. In this sense, average molecules representing the aromatic fraction contained in both GOV and tar pitches are proposed and a thermodynamic model at molecular scale is proposed to establish the ideal optimal conditions for the conversion of GOV into pitch, by means of a computational analysis. These conditions are then analysed after being evaluated in a batch reactor with a sand fluidised bed heating system.

The analysis and characterisation of the products will then be presented by means of the evaluation of physical and chemical properties and spectroscopic analysis, such as: softening point, density, viscosity, coking value (carbon residue), elemental analysis; saturates, aromatics, resins and asphaltenes, (S.A.R.A); gas chromatography (analysis of discriminated aromatics) and vapour pressure osmometry (V.P.O.).

CHAPTER 1

BASIC CONCEPTS FOR THE PITCH PRODUCTION

1.1. Oil

The word petroleum comes from the Latin words "PETRA" and "OLEUM", which means stone and oil, given that it is found in nature trapped in the rocks of the subsoil. The precise origin of petroleum is still unknown, but there is a widely accepted hypothesis, which states that petroleum is of organic and sedimentary origin, the product of the decomposition of matter from plant and animal remains deposited in seas and oceans, which, when covered by sediments, generate a complex physical-chemical process where pressure, heat, the presence of microorganisms and, above all, the passage of time, transformed these remains of materials into oils and gas within a porous medium normally composed of clays and sands, called petroleum source rock [i].

Petroleum is a natural, flammable oily mixture, ranging in colour from yellow to black, with a density equal to or less than water and, depending on its origin, can have a wide range of viscosities. In its natural state, petroleum consists of a complete series of solid, liquid and gaseous hydrocarbons.

These hydrocarbons are molecules consisting of carbon and hydrogen atoms, which vary in both the carbon-to-hydrogen ratio and molecular structure. At the same time, they possess in smaller proportions inorganic elements such as: oxygen (O), sulphur (S), nitrogen (N), vanadium (V), iron (Fe), nickel (Ni); in the form of functional groups such as phenols, cresols, mercaptans, thiophenes, sulphides, disulphides and organo-metallic compounds, which influence the quality of petroleum derivatives [i]. The typical percentage composition of petroleum as shown in Figure 1 is carbon 85%, hydrogen 11.98% oxygen 2%, sulphur 1%, vanadium 0.0075%, nickel 0.005%, iron 0.004%, copper 0.003% and other 0.0005% [ii,iii].

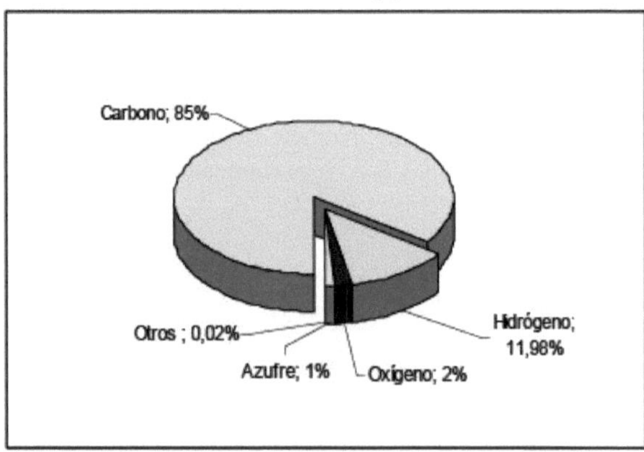

Figure 1. Chemical Composition of Petroleum [iii].

On the other hand, structurally, oil consists mainly of aliphatic, naphthenic and aromatic hydrocarbons. Often the percentages of these elements depend on the maturity of the reservoir. The most commonly used oil classification criteria are gravimetric (Table 1) and chemical. The first of these takes into consideration the API, American Petroleum Institute gravity (relative density) [i], which is calculated by the equation, described below:

$$Grados\ API = \frac{141{,}5}{\text{Densidad relativa}} - 131{,}5 \quad (1)$$

API Degrees
Relative density
Where the relative density is given by

$$\text{Densidad relativa} = \frac{\text{Densidad del líquido}}{\text{Densidad del agua}} \ (\text{A } 60°\text{F}) \quad (2)$$

Relative density
density
Water density

According to API gravity, oil can be classified as follows:

Table 1. Type of oil by density.

Oil or Crude	Density (g/ cm^3)	Relative Density (°API)
Extra-heavy	>1,00	<10,00
Heavy	1,00 - 0,92	10,00 - 22,3
Medium	0,92 - 0,87	22,30 - 31,10
Slight	0,87 - 0,83	31,10 - 39,00
Superlight	< 0,83	> 39,00

1.2. Refining

The set of physical and chemical processes that transform crude oil into useful products is called refining. Refining is the use of heat, pressure and/or chemicals to separate and combine the basic types of hydrocarbon molecules naturally occurring in petroleum into groups of similar molecules. Through specialised refining processes, the structures and bond types of the base compounds can be rearranged to convert them into products of higher commercial value and application [iv, v]. In most cases, the properties of the products resulting from the refining of petroleum and its fractions are directly related to the characteristics of the hydrocarbon to be separated or converted. Therefore, the most significant factor in the process is not necessarily the type of chemical compound used as a separation or reaction medium, but the type of hydrocarbon present in the crude oil (paraffinic, naphthenic or aromatic).

In general terms, three basic processes or operations can be distinguished in oil refining: separation processes, conversion processes and hydrotreating or purification processes.

The separation process consists of splitting the crude oil into different fractions without altering the basic structure of the compound. It is based on the difference in solubility, molecular weight, boiling point and/or polarity of the different fractions that make up the crude oil.

The conversion processes are based on the transformation of the molecular structure of the oil components, generally by the action of heat and/or with the use of catalysts. Hydrotreating or purification processes use hydrogen as the main input, which reacts and removes the sulphur, nitrogen, metals and oxygen compounds present in the hydrocarbons. This operation is carried out in order to meet commercial fuel specifications, which are closely related to the environmental aspect. Additional reactions occur in this process to convert olefins into saturated compounds and to reduce the aromatic content. Hydrotreating requires high pressures and temperatures, and the conversion is carried out in a chemical reactor with a solid catalyst consisting of gg-alumina impregnated with molybdenum, nickel and/or cobalt. For the scope of this research, the refining processes will be distillation and thermal cracking as methods used for the separation and molecular conversion of crude oil and its derivatives [ii]. It is important to note that, in refining, separation processes can be applied directly to crude oil or any of its derivatives, while conversion technologies are practically limited to the transformation of residual streams and fractions of crude oil, such as gas oils.

1.2.1. Fractional distillation

After the desalting of the oil, which consists of the removal of the small quantities of inorganic salts, which are generally dissolved in the remaining water, by adding a stream of fresh water (with low salt content) to the dehydrated oil stream, distillation is the first process that appears in a classical refining scheme. It consists of vaporising a liquid or mixture of liquids, condensing the vapour and collecting the component(s) by concentrating them in another vessel.

Fractional distillation is the process par excellence used in oil refining to separate its various components. In order for the separation of each of the components to occur, equilibrium must be reached between the liquid and vapour phases, since in this way the lighter or lower molecular weight components are concentrated in the vapour phase, and, on the contrary, those of higher molecular weight predominate in the liquid phase. In other words, distillation is based on the difference in volatility of the hydrocarbons present in the mixture (boiling point) [ii, vi, vii], these components are called fractions and are obtained in the form of liquid side streams when the oil is distilled in a column, called a fractionation tower. The column is kept very hot at the bottom and the temperature gradually drops towards the top, where the vapours pass into a cooling system called plates, in which the different fractions are gradually condensed according to their specific boiling points. Based on volatility, the fractions can be classified in descending order into gases, light distillates, middle distillates, gas oils and residues. Each of the fractions is stored and then fed to units that aim to recover the hydrocarbons.

1.2.2. Atmospheric and Vacuum Distillation

The classic refining scheme starts with a desalter, equipment that allows the removal of most of the salts and water contained in the hydrocarbon. The industrially desalted oil is preheated using heat recovered from the process, then transferred to a crude oil heater by direct heating and from there to a tower or vertical distillation column operating at temperatures around 340 °C and atmospheric pressure. In this equipment, physical separation into different direct distillation fractions is achieved, which include liquefied petroleum gas, naphtha, kerosene, light gas oil, heavy gas oil and a residue that does not evaporate, since it boils above 340 °C [viii, ix].

Since most of the hydrocarbons contained in the residue from atmospheric distillation start to react above 350 °C temperature (which alters the molecular structure of the hydrocarbon to be separated) it is pumped to a second column operating under vacuum conditions providing the reduced pressure necessary to avoid thermal cracking.

1.3. Cracking

After distillation, the different cuts or fractions of the crude oil are subjected to further treatments designed to modify the molecular structures of the feed fractions and increase the yield of light products. These systems are referred to as conversion processes. Among them is cracking, a reaction that essentially consists of breaking down or decomposing the long molecules of high molecular weight, high boiling point hydrocarbons into lower molecular weight hydrocarbons. This spontaneous reaction has a high activation energy, which means that the supply of heat and/or the presence of catalysts is necessary for it to occur. The usefulness of cracking lies in the high commercial value of the lighter petroleum fractions, mainly when they are used as fuels and base materials for further processing [x]. Cracking processes are classified into two types depending on the presence or absence of catalysts. The first is known as thermal cracking, where the energy required to break the chemical bonds is provided by an external heat source. On the other hand, catalytic cracking, in addition to heat, employs one or more chemical substances called catalysts that intervene in the speed of the reaction, accelerating or delaying to a certain extent the chemical reaction that takes place without modifying its mass [viii], contributing to the increase of selectivity towards the product, favouring the generation of the chemical compound that is required to be obtained.

1.3.1. Thermal Cracking

Thermal cracking phenomena occur as a consequence of the application of high temperatures (> 350 °C or 660 °F) in order to break, rearrange or combine hydrocarbon molecules without the intervention of catalysts [xi]. Thermal cracking of hydrocarbons has proved to be a very important and widely used industrial process. In this process, the heat of reaction required by the system increases as the boiling point of the feed increases. The cracking reactions are endothermic (heat/energy consuming), spontaneous and the disintegration of hydrocarbons is not selective (so that a variety of structures with different molecular weights and boiling points can be generated) because secondary reactions of dehydrogenation, condensation and polymerisation, which are exothermic, occur simultaneously [xii]. These thermal cracking reactions involving the formation of free radicals are responsible for generating lighter products, as well as producing heavier residues with a higher carbon-to-hydrogen ratio (C/H) than the feed, limiting the levels of thermal cracking by the stability of the product generated, this stability depends on the amount of high molecular weight chemical compounds formed and it is generally found that the higher the conversion, the higher the rate of production of unwanted by-products [xi, xiii, xiv, xv]. This is why, at the same time that thermal breakdown of heavy hydrocarbon molecules occurs, breaking them down into lighter fractions, reactions can occur that favour the production of coke as an unwanted by-product [xvi, xvii]. The key variables in thermal conversion operations are reaction temperature and residence time. The combination of these factors defines the severity of the treatment. Usually, this qualitative term refers to the degree of conversion obtained in the thermal decomposition of the charge [xviii]. However, although pressure is not such an important variable when setting these parameters in pyrolysis reactions, it is a factor that can influence the physical state of matter (gas,
liquid or solid), as well as on the chemical stability of the free radicals formed.

1.3.2. Thermal Cracking Chemistry

Due to the multiplicity of reactions that occur during thermal cracking of hydrocarbons from crude oil processing, the representation of a mechanism cannot be expressed in a simple chemical equation. However, this type of reaction occurs at high temperature, and is based on a generalised mechanism developed by Kossiskoff and Rice [xix, xx], which justifies that thermal decomposition involves a chain reaction of free radicals (chemical compounds possessing an unpaired electron). The reaction steps include the following stages:

- Initiation: introduction of the free radical into the reaction system. In pyrolysis the reaction is initiated by an external agent, generally this thermal treatment is of a severe type.

$$R_n \xrightarrow{k_1} 2R^{\bullet} \qquad (3)$$

- Propagation: series of reactions that convert reactants into products, while leaving the radical concentration unchanged. Typically, propagation steps include: hydrogen transfer, isomerisation, radical decomposition (its reverse reaction) and radical addition.

Chain transfer: $R^{\bullet} + R_n \xrightarrow{k_2} RH + R_n^{\bullet}$ (4)

β - Incision: $R_n^{\bullet} \xrightarrow{k_3} R^{\bullet} + Olefina$ (5)

- Termination: combination and/or disproportionation of radicals to give stable products. In this process, one of the radicals transfers a hydrogen atom to the other radical to produce an alkane and an alkene.

$$Radical(R_n^{\bullet} o R^{\bullet}) + Radical(R_n^{\bullet} o R^{\bullet}) \xrightarrow{k_4} Productos \qquad (6)$$

Equation (5) is called β-Initiation, because the C-C bond is located two atoms away from the hydrogen-deficient carbon where bond-breaking occurs to form an olefin and a smaller alkyl radical. Chain reactions occur when propagation reactions (4) and (5) are more frequent than initiation (3) and termination (6), because free radicals accumulate to a pseudo-steady state that allows thermal cracking reactions. An example of reactions via β-initiation is observed with the n-butane radical, which is presented below in equations (7) and (8):

$$CH_3CH_2CH_2CH_3 + R^{\bullet} \longrightarrow RH + {}^{\bullet}CH_2CH_2CH_2CH_3 \qquad (7)$$

$${}^{\bullet}CH_2CH_2CH_2CH_3 \longrightarrow CH_2 = CH_2 + {}^{\bullet}CH_2CH_3 \ldots \qquad (8)$$

The reactivity of the feed components determines the reactions of thermal cracking. Thus, the chemical structural reactivity of the families will increase in the increasing order of paraffin molecules, naphthalenes, olefins and aromatics (Figure 2).

The bond dissociation energies are presented in Table 2, which provides an overview of the difficulty of breaking the different types of bonds present in the hydrocarbons that make up the crude oil [xxi].

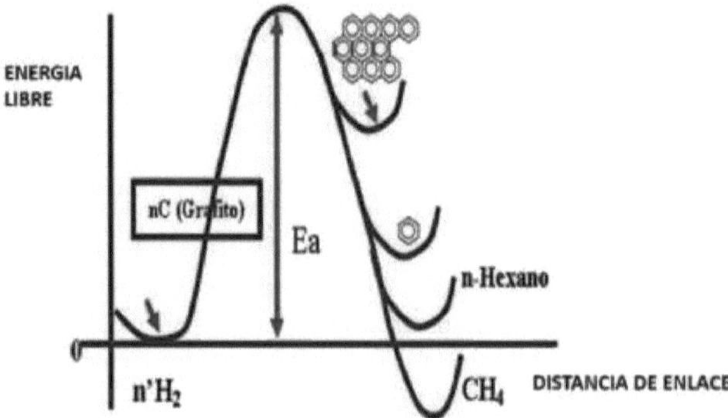

Figure 2. Free energy Vs. residence time.

Table 2. Binding dissociation energy [xiv, xv].

Link	Binding dissociation energy (Kcal. /mole)
C-C	*82,6*
C=C	145,8
C C	199,9
C-H (n-alkanes)	98,7
C-H (aromatics)	110,5
H-H	104,2
C-O (methoxy)	85,5
C-S	65,0
S-S	84,0
S-H	83,0
C-N (amine)	72,8
C=N	147,0

From Table 2 it is noted that the dissociation energy of the carbon heteroatom bonds is much lower than that of other bonds typical of structures from petroleum fractions. For this reason, the breaking of these types of bonds will occur first, which implies that the higher the content of heteroatoms such as sulphur and nitrogen among others, within the molecular structures of the filler, the higher the rate of breaking of the molecules via thermal cracking.

Table 3 presents the variety of reactions that occur depending on the type of feed or charge, as well as the type of functional group they generate after thermal decomposition.

Table 3. Most likely thermal cracking reactions for various types of hydrocarbons [xxii, xxiii].

hydrocarbons [xxii, xxiii].

Type of thermal cracking reactions most likely to occur for compound various types of hydrocarbons

Decomposition, forms another paraffin and an olefin.

Paraffin waxes Dehydrogenation, forms an olefin from the same length of the chain.

Alkylation, cyclisation and isomerisation can occur at high temperatures.

occur at high temperatures.

Partial decomposition, formation of other olefins, paraffins and dienes.

Dehydrogenation, producing dienes.

Polymerisation

Olefins

Secondary reactions between olefins and dienes, can generate cyclic olefins.

Isomerisation, to produce naphthenes.

Dehydrogenation of naphthenes and cyclo-olefins to produce aromatics.

to produce aromatics.

Dealkylation, losing paraffinic chains Naphthenes side chains, forming paraffins, olefins and dealkylated naphthenes with or without substitutions.

Dehydrogenation producing aromatics.

1.3.3. Catalytic Cracking

The heavy fractions are disintegrated in the presence of a catalyst, a substance that is present in the system under study in physical contact with the reactants and is capable of generating changes without altering the final energy balance of the chemical reaction [xxiv]. The catalysts used in this type of cracking modify the reaction rate. Generally in the oil industry, they aim to accelerate the bond-breaking processes, while maintaining the selectivity of the reactions.

There are three basic stages in catalytic cracking:

- Reaction: the charge reacts with the catalyst and decomposes into different streams.
- Regeneration: the catalyst is reactivated by burning coke.
- Fractionation: the cracked hydrocarbons are separated into various products for downstream processes.

1.4. Thermal Cracking Processes in Refining

After distillation, most of the separated products are redistilled again for purification. The residues are vacuum fractionated to obtain the feedstock required in the production of lubricating oils, diesel, kerosene, jet fuel, petroleum coke, asphalt and fuel oil among others, while other by-products are used as feed to downstream processes. However, distillation processes rarely yield products with the quality and volume demanded by the market. For this reason, modern refineries use conversion technologies to obtain the products needed by consumers. Among these processes are those that break down large molecules, usually present in the heavy fractions, to form smaller molecules of more commercial application. Industrial refining processes using thermal cracking for the valorisation of crude oils and waste streams include: visbreaking, production of oil tar pitches and mesophase-rich pitches, fluid coking, flexicooking and delayed coking [xxv, xxvi].

1.4.1. Viscoreduction

It consists of reducing the viscosity of both atmospheric and vacuum residue by means of thermal cracking reactions of moderate severity, with the purpose of refining its use as a fuel, as well as increasing its handling and transfer through pipelines.

The process involves a mechanism based on the breaking of carbon-carbon bonds that allow the release of aromatic polycondensed systems, producing free radicals that generate chain reactions, responsible for considerably reducing the activation energies of the process. The reactions of rupture in the beta position with respect to the free radical generate olefins and release hydrogen, which subsequently saturates the free radicals formed [xxvii, xxviii].

Its main product is fuel oil, a mixture of residues that needs to be diluted with lighter by-products of higher value, which can be sent to other conversion processes and thus obtain a higher added value. One of its main objectives is to reduce the consumption of diluent that is mixed with the residual fuel in order to achieve the specifications of the product.

Viscoreduction is one of the simplest and most economical processes used for the reduction of residuals and increasing the yield percentages of distillates. Currently, two variants of this process are known [xxix].

- Visco-reduction with furnace, at operating conditions of high temperatures in the range 480 - 490 °C and low residence times (1 to 2 minutes).
- Soaker visbreaking: working conditions in the temperature range of 440 to 460 °C and reaction time of 10 to 20 min. This method focuses on reducing the viscosity of liquids, which can be essential in manufacturing and production processes.

Soaker" visco-reduction is characterised by the use of smaller furnaces that are less sensitive to operational changes, lowering energy costs
(30-35%) and increase the reaction rate. The residue from this process is very unstable, usually used as feedstock for the production of residual fuel (fuel oil).
The range of finished compounds derived from visbreaking includes: Light gases used in liquefied petroleum gas or as sources of olefins for alkylation, vacuum gas oil for catalytic cracking processes, olefin-rich gasolines and gas oil with high aromatic content.

1.4.2. Delayed Coking

Thermal conversion process based on breaking the carbon-carbon and carbon-hydrogen bonds that make up the long molecular chains of heavy and extra-heavy hydrocarbons. It was originally developed to minimise the yield of heavy residual fuel in refineries by applying a severe thermal treatment (pyrolysis), yielding coke as one of the end products. In addition to coke, fuel gas, unstabilised naphtha and other distillates (gas oils) used as feedstock in hydrodesulphurisation processes are produced.
The chemistry of the system is summarised in a three-stage thermal cracking reaction based on free radical formation, polymerisation and condensation. The furnace provides the heat necessary to generate the cracking by keeping the reaction in the coke drum. This process generally uses residues from vacuum distillation, feed which is heated for a short period to over 490 °C and pumped into one of the two coke drums.
In delayed coking [xxx], the coke drums operate as reactors where the remaining reaction stages are completed, producing vapours which are released from the top and pass to the bottom of the fractionator. The highly viscous intermediate is held in the reactor and is progressively transformed into a carbonaceous material called coke. The operating conditions must be strictly controlled to minimise undesirable effects such as coking in the oven tubes. From the point of view of the endothermic reactions occurring in this process, three main stages follow: main vaporisation and relatively mild cracking in the oven, cracking of the vapours in the drum and cracking with polymerisation of the coking mass in the drum.
In addition to the cracking reactions, side reactions occur between carbon atoms and inorganic atoms or heteroatoms (nitrogen and sulphur), which can negatively interfere with downstream refining processes, e.g. by poisoning the hydrotreating catalysts.

1.4.3. Fluid Coking

A continuous process that allows heavy hydrocarbons to be converted into light products with the same levels of inversion as those achieved with delayed coking. Essentially, it is a non-catalytic thermal conversion through free radicals, using as feedstock, atmospheric residue, vacuum residue and deasphalted bottoms.
As in delayed coking and other coal reject technologies, a significant proportion of the contaminants present in the feed (metals, coke, sulphur, nitrogen) are deposited in the coke.
Some typical values for the rejection of these pollutants are summarised in Table 4. The main equipment of the process is a fluidised bed reactor and burner. An integral part of the thermal conversion reactor is the quench or injection of a low-temperature scrubber product into the reactor vapour outlet stream in order to rapidly stop the thermal conversion reaction.

Table 4. Contaminant rejection [xxxi].

Liquid rejection (%of feed)

Feeding	Sulphur	Nitrogen	Nickel	Vanadium
Arab Heavy	32	97	95	95

Cold Lake Bitumen	34	79	99+	99+
Hondo	24	81	99+	99+

Fluid coking can be operated in one of two modes, with conventional or single-step recycle. In the recycle mode all material at temperatures above 524 °C is converted into light products and coke. Typically, the feed is preheated to 260 °C then fed to the scrubber to be preheated with the reactor's top steam stream, the scrubber vessel or pool reaches temperatures in the order of 372 °C by integration of the heat from the cooler feed with the reactor product gases. In this way, the heavier hydrocarbons in the product stream are condensed and co-mixed with the feed that will ultimately be recycled in the reactor. On the other hand, in the once-through mode, material at temperatures above 525 °C is removed from the scrubber to remain part of the heavy gas oil product. A variation of the process is the integrated vacuum "Pipestill" mode , where the cut-off point of the recycled vapours increases from 525 °C to 565 °C [vii]. In the reactor, the temperature range varies between 525 and 565 °C, the heavy hydrocarbon feed (both fresh feed and fresh feed/recycle mixture) is injected into the fluidised bed of coke fines by means of a series of injectors located in rings around the vessel. Several of these rings are oriented at various levels in the vessel. By spraying onto the hot coke bed.

The feed is thermally cracked into a full range of lighter products, from gases to gas oils, yielding coke as a by-product. The hydrocarbon vapours pass through the dense and dilute phase of the fluidised bed to the system at the top of the reactor, which consists of a set of cyclones for the recovery of the coke fines, which are returned to the dense bed, and to the reactor scrubber. The coke fines pass through the cyclones and are captured in the scrubber. Depending on the mode of operation, these fines will be recycled to the reactor with the +525 °C fraction, or an alternative removal system will be needed to protect the downstream equipment. The product flows from the scrubber to the conventional gas pressure fractionator and to the fines recovery units. Conventional fractionation facilities used to separate liquid products may incorporate their own light plants or may be integrated with another process unit.

Coke produced by thermal cracking reactions is deposited in the bed of coke fines, typically in layers. In addition to the feed, steam is introduced, some of it with the feed and some separately. This high temperature steam provides additional stripping of hydrocarbon vapours and coke fines to maximise the production of coke fines. The steam introduced from the bottom of the reactor also serves to fluidise the coke particles. The total coke inventory is maintained in the reactor by removing the bottom coke, through the removable section, transferring the surplus to the burner and by circulating hot coke to the burner where the net coke is discharged. The heat required for the endothermic reactions of the process is supplied by recirculation of hot coke from the burner to the reactor.

1.4.4. Flexicoquification

Extension of the fluidised coking process that includes coke gasification within the scheme. This is achieved by incorporating a third fluid solids vessel into the flow scheme to gasify up to 97% of the coke produced in the unit, and includes desulphurisation of the low BTU (British Thermal Unit) gas. While the overall liquid outputs are comparable, gasification of the coke produces a combustible, low BTU gas called sweet LBG suitable for use within the plant or at neighbouring facilities.

Since the coke production is converted to a clean fuel gas, flexicooking maximises the total hydrocarbon yield of the plant, through the substitution of low BTU gas for plant fuels. As an extension of fluid coking, the process elements in the reactor are the same as in the reactor, including the cold coke and hot coke transfer lines. The nomenclature for the burner vessel is changed to a heater vessel. The gasifier is the third vessel that is added to the fluid coking process to gasify the coke with steam and air, producing syngas (H_2, CO, CO_2 and inerts). The operating temperature for the gasifier is approximately 925 °C and 980 °C. The gas produced together with the entrained particles, , is routed through the heater vessel for fluidisation of the hot coke bed and to transfer heat to the solids. The heat balance for the reactor is achieved by circulating the heated coke and the gasified coke. The gas is removed from the heater through a cyclone system and additional heat is removed through a steam generation train. Downstream in the heat recovery system is a third cyclone stage followed by a Venturi scrubber to recover the remaining coke particles. The low BTU gas, or syngas, is then treated for hydrogen sulphide removal. Typically, a small amount of the coke fines (approximately 3%) is recovered at the heater top by purging the metals from the heater and gas generator. Another fraction of the additional coke is removed from the heater. The metal compounds present in the cokes produced by this process are mainly nickel and vanadium [xxxii].

1.4.5. Oil Tar Blast Production

Moderate thermal cracking process of raw materials, generally highly aromatic fillers or atmospheric and vacuum residues. In order to promote polycondensation reactions leading to the production of pitches with physico-chemical properties suitable for use as a binder for petroleum coke used in the production of anodes consumed in the aluminium industry as graphite electrodes. As in the case of other thermal cracking processes, polycondensation, condensation and dehydrogenation reactions occur. Coal tar pitch production technologies are an alternative to the use of coal tar pitches, a product obtained from the pyrolysis of coal or metallurgical coke. In oil tar pitch production technologies, the feed is pre-heated, then pumped into a soaker reactor or stirred tank with progressive heating, to be treated under controlled conditions of pressure and temperature to promote condensation and polymerisation reactions.

The treated fraction is transferred to a fractionation or distillation tower where it is separated into gases, light distillate and a bottom stream or fraction typically referred to as oil tar pitch base. This base is subjected to vacuum distillation in order to concentrate in the bottom product the polyaromatic compounds that impart the binding properties of petroleum tar pitch. These properties can be enhanced by using additives such as carbon black, finely divided coke (coke fines), light gas oil or a mixture of these [xxxiii]. Within the process, the fresh feed is added to a portion of the bottom stream, which is recirculated (called recycle stream). The amount of the recycle stream added to the fresh feed allows a control of the quality of the developed petroleum pitch. In addition, the heating, reactor reactions and fractionation stages are conducted in an oxygen (O_2)-free environment, specifically in an inert environment such as argon, nitrogen or similar. It has been found that, by conducting the technical process in an oxygen-free environment, an increase in the quality and yield of petroleum tar pitch can be achieved. The cargo or feed, whose characteristic is high aromaticity, is kept in the tank (A), extracted by the pump (B) and heated in a furnace (C) in a temperature range of 300 to 330 °C, taking care to avoid any change in the chemical composition. Then the cargo is transferred to a reactor whose temperature is adjusted between 370 °C and 460 °C, where thermal cracking of the cargo takes place for a time of less than 60 minutes and pressure of 10 atmospheres absolute [xxxiv]. Finally, the distillates are circulated to the separation tower (D), while the residue is finally pumped to the distillation unit (E) where the finished product is obtained.

1.4.6. Mesophase Breas Production

Mesophase pitches are materials with the characteristics of a liquid crystal, consisting of sphere-shaped basic units with a structure similar to that of graphite, although the way in which the graphitic planes are stacked is very different from that of graphite. To obtain mesophase pitches and mesophase fibres, the highly aromatic material is converted into pitch with suitable characteristics, then heat treated at around 400 to 500 °C [xxxv] with different retention times depending on the material used to obtain the mesophase pitch. In the next stage of the carbon fibre process, the pitch is spun into a thermoplastic fibre, with the molecules oriented along the axial axis. The extruded material is then oxidised to produce a thermosetting mesophase pitch fibre, with oxygen cross-links between the molecules maintaining and locking the orientation. Finally, they undergo a heat treatment process in inert gas, the temperature of which can vary in the range of 1500 °C to 3000 °C, to produce carbon or graphite fibres, respectively.

Carbonaceous mesophase is the result of the structural transformation of a liquid state, in which high molecular weight aromatic molecules are produced by pyrolysis reactions to form a parallel alignment of anisotropic liquid crystals [xxxvi].

1.5. Petroleum Tar Breas (BAP)

A highly aromatic petroleum-derived material used as a binder for solid aggregates in the production of anodes for the aluminium industry and which replaces coal tar pitch produced through the pyrolysis of coal. Their characteristics include: their boiling range for obtaining tar from coal tar is between 450 °C and 500 °C. In addition, they exhibit a wide softening range rather than a defined melting temperature and their ratio of aromatic hydrogen to total hydrogen varies between 0,3 and 0,6. Aliphatic hydrogen atoms are mainly contributed by alkyl-type groups substituted on aromatic rings or naphthenic rings [xxxvii].

The industrial process of obtaining aluminium is based on the electrolytic reduction of aluminium oxide (Al2O3) or alumina, dissolved in a bath of cryolite (double fluoride of aluminium and sodium) melted at 1800 °C, to which a strong electric current is passed. The metallic aluminium is separated from the chemical solution and extracted by means of an inverted U-shaped tube, one end of which is immersed in a liquid which rises up the tube higher than its surface and drains out at the other end. In this cryolite bath, under the effects of the electric current, the alumina is converted into aluminium in the proportion of one kilogram of aluminium for every two kilograms of alumina with the intervention of carbon anodes.

This process of obtaining aluminium is known as the Hall-Heroult Process, whose inner surface of the cell is coated with carbon molecules and carbonated iron which functions as a cathode, and in which aluminium ions are reduced to the free metal. The reaction describing the system is shown at
continuation:

Cathode $$4\left[Al^{+3} + 3e^{-} \xrightarrow{Electricidd} Al(l)\right] \quad (8)$$

Anod $$3\left[C(s) + 2O^{-2} \xrightarrow{Electricidad} CO_2(g) + 4e^{-}\right] \quad (9)$$

General reaction

$$2AlO_3(sol.) + 3C(s) \xrightarrow{Electricidad} Al(sol.) + 3CO_2(g) \qquad (10)$$

The anodes in equation (9) are typically composed of 65 % petroleum coke, 15 % coal tar pitch and 20 % capes (waste anodes whose origin can be: anodes consumed in the process of obtaining aluminium, green or raw anodes that did not pass the baking and rodding processes or defective baked anodes). These anodes are progressively consumed in the 1:2 carbon-aluminium reaction and are transformed into gaseous co2 and elemental aluminium, which is why they must be replaced frequently, with the consequence that this raw material represents one of the main costs associated with the production of aluminium. Although this reaction is the one that governs the process, there are others, such as the following:

$$2Al(sol.) + 3CO_2(g) \longrightarrow Al_2O_3(sol.) + 3CO \qquad (11)$$

This reaction is responsible for decreasing the efficiency of the metal stream, since when CO is formed, the total consumption of carbon per unit of metal produced increases, which is detrimental to the economy of the process. As could be seen from these reactions, the aluminium production process is closely related to the quality of the carbonaceous materials (coke and pitch) used to manufacture the anode and the affinity that must exist between them. Thus, an out-of-specification tar pitch can lead to anode fractures due to thermal shock effects, highly porous and not very dense anodes, and therefore highly reactive to air and co2, which increases the net carbon consumption. Similar behaviour is to be expected when using cokes with high coefficients of thermal expansion and high concentrations of oxidation catalysts and co2 reactivity, such as sodium, nickel and vanadium. The feedstock for petroleum pitch production is a highly aromatic organic material (tar), a black, highly viscous and dense solid residue at room temperature, derived from the thermal cracking of these materials and consisting of a mixture of numerous predominantly aromatic and alkyl substituted hydrocarbons. It is of relatively low economic value, however, they are the precursors to products of high commercial value, such as advanced carbon and carbon fibre or graphite materials. The conditions of this process and the characteristics of the precursor will determine the properties of the resulting carbon material.

Oil tar pitches can be obtained from various sources, including: from the bottom of the atmospheric and vacuum tower, in catalytic cracking processes, from steam-cracked pitch, as a by-product in the treatment of naphtha, or any other refinery residue; however, most of them are produced today as a by-product of the coke produced by the industry. Various studies carried out by PDVSA have demonstrated the technical and economic feasibility of replacing coal tar or coal tar pitches with petroleum tar pitches, as well as their importance in terms of environmental impact, as they have shown lower emissions of polyaromatic hydrocarbons (PAHs) during the anode manufacturing process.

Petroleum coke is a solid material obtained as a by-product of high severity thermal cracking reactions or carbonisation processes. It is an infusible solid and generally anisotropic (directionally dependent). It can be obtained from hydrocarbons with low resin content and partially or fully aromatic or heterocyclic asphaltenes, derived from petroleum or hard coal. Given its porosity and black to bright grey colour, it is indistinguishable at first glance from coke obtained from coal, especially in the case of delayed coke. The characteristics of tars and pitches are dependent on the type of material used, the operating conditions of the production ovens, the operating temperatures, the reaction time and the charging method, the latter having a great influence on the quality of these products [xxxviii]. Within the specifications of petroleum tar pitches as raw material (binder) in the production of carbon anodes for the aluminium industry are listed in Table 5.

Table 5. Specifications for oil tar pitches.

Property	Specification Range
Softening point, °C	125,0-130,0
Viscosity at 160 °C, cP	<10000,0
Coking value, % w/w	>48,0
Insoluble in toluene; % w/w	>9,0
Quinoline insoluble; % w/w	0,6
Density, Kg/dm^3	1,2

1.5.1. Chemistry

The composition of petroleum tar pitches varies considerably depending on the characteristics of the filler used for their production, as well as on the reaction time and the heat treatment applied. From these considerations it can be stated that pitches are generally made up of a mixture of hydrocarbons of complex chemical composition, but their constituents belong to a few classes of compound families. Despite the diversity of their constituents, this is compensated by the similarity of their families [xxxix]. Among the types of families of chemical compounds present in this type of material, the following can be mentioned: polycyclic aromatic hydrocarbons (alkyl substituted, with the cyclopentadiene group, partially hydrogenated, heterosubstituted, with carbonyl groups, among others), oligoaryls and ologoarylmethanes, polycyclic heteroaromatic compounds (pyrrole benzologues, furan, thiophene and pyridine).

Polycyclic aromatic hydrocarbons (PAH) are the most abundant group of compounds in pitches. Depending on their structure, they can be classified into cata-condensates and peri-condensates. Cata-condensates have tertiary carbon atoms common to at most two aromatic rings, whereas in peri-condensates, some tertiary carbon atoms belong to three aromatic units. The different typology of these two classes of polyaromatic compounds affects their behaviour, e.g. their thermal reactivity.

Using techniques such as extrography and laser desorption mass spectroscopy, molecular species of high weights, from 595 Daltons to around 12000 Daltons, have been found. Furthermore, by Tanden masses, it has been shown that partially alkylated diaryl-methanes are typical structures appearing in the middle molecular weight range of petroleum tar pitches [xl]. Likewise, by gel permeation chromatography (GPC), using quinoline as solvent, results were found for molecular weights in the range of 450 to 2000 Daltons. The heterocyclic portion of high molecular weight compounds of the pitch is estimated to represent 10-15 % of the relative weight of the pitch. Their ratio of aromatic hydrogen to total hydrogen varies between 0.3 and 0.6 [xli].

When comparing petroleum tar pitches with coal tar pitches (which are the result of coal distillation or pyrolysis) the molecular weight distribution is broader and the average masses are higher, the content of heterocyclic compounds is lower, particularly applied to nitrogen-containing heterocycles.

1.5.2. Characterisation

They are traditionally characterised by means of three methods, which must be combined to obtain the most accurate representation possible. These methods are given by physical properties, structural features and individual analysis of the pitch components.

Characterisation by physical properties involves techniques such as:

- Softening point: a test defined as the temperature at which tar changes from a brittle solid to a viscous liquid [xlii] .This definition is inaccurate, as there is no precise quantification of viscous liquid, but this concept is used because pitches when heated behave like glass-forming materials, so they do not undergo the classic solid-liquid phase transition as they are heated, but pass through a glass transition region before forming a viscous liquid [xliii], thus avoiding having an established melting point. This technique is generally the most widely used as a specification for pitches, as it is relatively easy to perform, and is also a valuable indication of other properties, such as coking value, insoluble content and density, which are directly dependent on it [i].
- Coking Value (Coal Residue): As described in petroleum tar pitch chemistry, these are mixtures of a wide variety of compounds that differ in their physical and chemical properties. A light portion of these compounds can be vaporised in the absence of air at atmospheric pressure without leaving an appreciable residue. Other non-volatile compounds leave a carbonaceous residue when a destructive distillation is carried out. This residue left behind is called carbon residue and represents a key property in determining the carryover of heavy components into products in atmospheric and vacuum distillation processes, and in predicting the yields and qualities of thermal conversion processes. This value is closely linked to the API gravity and the asphaltic content of the filler or product, as well as the aromaticity of the pitch and its binding properties.
- Insoluble Material Content: materials insoluble in certain solvents, which are present in the pitch and cannot be separated by distillation. They can be divided into two groups: those insoluble in toluene and those insoluble in quinoline. Toluene-insoluble hydrocarbons are the group of hydrocarbons that significantly improve the binding capacity of the pitches, since, due to their average viscosity, these compounds have the ability to penetrate the coke mixture and hold the petroleum coke particles together at an intramolecular level. In turn, these materials have a great influence on the viscosity, since the higher the concentration, the more constant the viscosity remains with increasing temperature.

Quinoline insolubles are particles consisting of arrays of spheres ranging in size from $1X10^{-6}$ to $4x10^{-6}$ metres [xxx]. It is this size that provides the ability to fill the intramolecular spaces in the mixture and displace the air trapped there , affecting important electrode properties such as electrical resistance and density. It is important to note that in petroleum tar pitches, these insolubles are not present, so in order to make up for these properties, other properties must be increased, mainly the carbon residue. The lack of quinoline insolubles results in a detrimental effect on the binding quality of the coke aggregate.

- Density: is defined as the mass per unit volume of a material. Relative density can be defined as the ratio of the mass of a given volume of a material to the mass of the same volume of water at the same temperature. A low density in the oil tar pitch also results in low density at the anode, higher porosity and higher carbon consumption in the aluminium reduction cell.
- Viscosity: Measurement of the resistance of a liquid to flow under pressure, generated by a mechanical source. It is based on measuring the time it takes for a fixed volume of liquid to flow under gravity through a calibrated glass capillary tube under standard conditions and at a fixed temperature. The behaviour of non-Newtonian liquids whose viscosity varies with the stress gradient applied to it (as a result, a non-Newtonian fluid does not have a defined and constant viscosity value, unlike a Newtonian fluid) can be determined using various rotational

viscometers, mainly cone-plane viscometers.

In pitches, the applied viscosity is the rotational viscosity, which influences the binder behaviour in the anode preparation process and the distribution of the viscosity in the mould [xliv]. A low rotational viscosity affects the overall porosity of the anode and increases the carbon consumption in the cell. High viscosity increases the possibility of mosaic formation, affecting the molecular rearrangement (mesophase formation), hence the molecular structure of the coke. The size and shape of the optical texture determine properties such as mechanical strength, reactivity and thermal resistance [xlv]. Lower viscosity will allow the use of mixing temperatures similar to or lower than those used for anode production, resulting in energy savings.

Regarding the characterisation of the structural features of the tars, the techniques involved that stand out are:

- Simulated distillation means the set of methods using the gas chromatographic technique for the determination of the boiling point range and distribution of hydrocarbons between -45 and 750 °C. Simulated distillation methods are based on the chromatographic separation of hydrocarbons in the order of their boiling points, when using a packed or impregnated column with a non-polar liquid phase and running a linear programme of the furnace temperature, allowing to obtain the percentage by weight of product existing in the temperature range of the furnace [xlvi]. On the other hand, this type of chromatography is used in a different way to the conventional one, which would be to achieve an optimal separation of the components of the mixture, as the working conditions are selected to obtain the separation with a limited resolution and efficiency [xlvii]. For this purpose, a calibration with n-paraffins of known boiling point is used. The resolution of the hydrocarbon components is the average of the boiling points of the mixture and is measured by the Flame Ionisation Detector (FID). FID offers high sensitivity for hydrocarbons, especially for high-boiling hydrocarbons, and low sensitivity for CS_2, which is why it is used as a solvent in sample preparation [xlviii], [xlix].
- Determination of Relative Molecular Mass (Molecular Weight): a fundamental physical constant that can be used in conjunction with other physical properties to characterise pure hydrocarbons and their mixtures. The commonly used method for determining the value of the average molecular weight in pitches is Vapor Pressure Osmometry (V.P.O.). This technique is based on determining the number average molecular mass (Mn) of a compound, which is given by the equation:

$$M_n = (\sum_i N_i M_i / \sum_i N_i) \qquad (12)$$

Where, Ni = number of moles or molecules with molecular mass Mi.

Vapour pressure, boiling point and osmotic pressure are colligative properties of a solution, which vary linearly as a function of the concentration of solute particles. When compared to a pure solvent, these properties will be affected proportionally to the number of solute particles dissolved in one kilogram of solvent. Therefore, measuring the change in one of the properties will indirectly give the value of the solution concentration or osmolality [l].

Osmolality is the total concentration of dissolved particles in a solution, regardless of properties such as particle size, density, configuration or electrical charge. The measurement of osmolality is based on the fact that the addition of solute particles to a solvent changes the free energy of the solvent molecules, transforming their colligative properties. When compared to the pure solvent, the vapour pressure and freezing point of a solution are lower in magnitude, while its boiling point is higher, provided that only one solvent is present in the solution. Thus, the relative changes in the properties of the solution are linearly related to the number of particles added to the solvent, but not necessarily related to the weight of the solute, since solute molecules can dissociate into two or more ionic components.

- Determination of Saturates, Aromatics, Resins and Asphaltenes (S.A.R.A) fractions: because the composition of oil is complex and has innumerable components, this analysis allows to check and compare its composition by separating it into four families of compounds: Saturates, Aromatics, Resins and Asphaltenes, called as SARA fragments. This is determined by the High Performance Liquid Chromatography (HPLC) technique, which involves the quantitative determination and separation of the saturated components, resins, aromatics and asphaltenes from the non-volatile crude oils using a High Performance Liquid Chromatography (HPLC). Classical" liquid chromatography is carried out on a column, usually glass, packed with the stationary phase. After seeding the sample at the top, the mobile phase is made to flow through the column by gravity. In order to increase the efficiency of the separations, the size of the stationary phase particles is decreased down to micron size, which generated the need to use high pressures to achieve the flow of the mobile phase. Thus, the technique of high-performance liquid chromatography was born, which requires special instrumentation to work with the high pressures required [li].

Finally, in the characterisation of the tars, the individual analysis of the components is presented, based on qualitative and quantitative determination:

- Determination of Carbon (C) and Hydrogen (H): the identification of these elements in organic compounds is based on the chromatographic analysis of the gases that are released from the sample after being subjected to high temperatures in an atmosphere of purified oxygen , where a combustion process occurs and generates carbon dioxide and water vapour among others, which are isolated for their respective quantitative determination. The values obtained represent the percentage of carbon and hydrogen. This method is applicable to oil samples from which the simultaneous determination of these two elements is carried out. This analysis is useful to determine the complex nature of the types of samples to be treated, the results obtained help to estimate the processing, refining and production potential in the petrochemical industry [lii].
- Nitrogen determination: chemiluminescence allows its quantification. It is based on the emission spectrum of an excited species formed in the course of a chemical reaction in some specific cases, where the excited particles are the product of the reaction between an analyte and a suitable reagent, usually a strong oxidant such as ozone, resulting in an emission spectrum characteristic of the oxidation product of the analyte or reagent rather than the spectrum of the analyte itself, the resulting signal is a direct, reliable and accurate measure of the nitrogen content of the sample [liii].
- Sulphur determination: the method used is X-ray fluorescence. It consists of the absorption of x-rays by an atom in the sample and its subsequent ejection of base or valence electrons. This method provides information about the elements present in the extract analysed, as well as the binding energies of the ejected electrons and the changes relative to these energies [liv].

The background intensity measured at a recommended wavelength (5 190 Â) is subtracted from the maximum intensity. The difference between the two signals is proportional to the sulphur content of the sample. This result is compared on a previously prepared calibration curve, to obtain through an equation the total sulphur concentration in percent (%) [lv].

- Determination of metals: this is carried out by means of the atomic emission spectroscopy technique based on the detection and quantification of the light emitted by an atom, molecule or ion, which has been subjected to a previous excitation process in a stream of ionised gas at high temperature (plasma).

When a set of atoms is subjected to very high temperatures, a large number of these atoms are excited and emit light when they return from one excited level to a lower energy level, the intensity of the light emitted is proportional to the total population of atoms in the higher energy excited state.

1.5.3. Average Molecular Structure

Many of the main properties of pitches, particularly the rheological ones, as well as the thermochemical behaviour, depend basically on the average molecular weight and the distribution of these in all components, which implies that their structural characteristics are directly related to the physical and chemical behaviour. In the following, some typical average molecular structures of BAP petroleum tar pitches reported in the literature are presented (Figure 3):

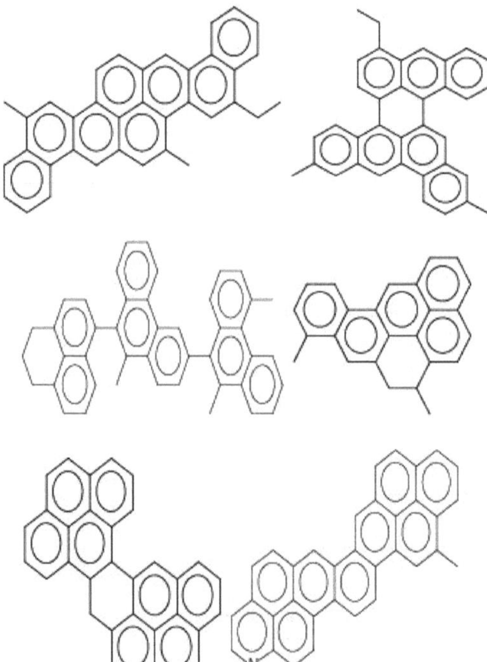

Figure 3. Average molecular structures of petroleum tar pitches [lvi, lvii].

As can be seen, the structures presented are generally molecules whose mass ranges between 600 and 12000 Dalton, with minimal heteroatom content, as well as having naphthenic structures and aliphatic chains that do not exceed two carbons. They consist of aromatic nuclei of 5 to 9 rings linked together by aliphatic and/or naphthenic bridges, distributed in both peri- and cat-condensed arrangements in the same average molecule with a high preference for cat-condensed structures.

1.6. Vacuum Gas Oils

In the early 20th century, the fraction collected at temperatures between 200 °C and 400 °C, which was used to produce town gas for lighting, was called vacuum gas oil. Today this name is used for intermediate oil fractions (heavier than naphtha and lighter than fuels), obtained in the vacuum distillation process, generally used as feedstock for secondary processes such as thermal cracking to obtain and produce liquefied petroleum gas (LPG), gasoline and other derivatives.

Within certain limits, it can be used as a diesel fuel or as a diluent for other fuel oils [lviii]. Depending on the nature of the crude oil, vacuum gas oils have different percentages of saturates, aromatics, resins and asphaltenes, the latter in very small, usually trace proportions. The separation takes place inside a vacuum column, obtaining at the top of the unit a fraction with a boiling temperature range between approximately 350 °C and 500°C free of the heavier components which remain in the fraction called residue [lix].

1.6.1. Chemistry

Vacuum gas oils are a mixture of numerous saturated hydrocarbons, aromatics, resins, asphaltenes and heterocyclic compounds that may contain metals. About 92 to 97% of the weight of vacuum gas oils consists of carbon and hydrogen, which is called hydrocarbon. The remaining portion consists of two types of atoms: metallic and diatomic. Diatomic molecules, such as oxygen, nitrogen or sulphur, often replace carbon atoms in the average molecular structures of these fractions. These characteristics define many of the chemical and physical properties of gas oils. The type and quantity of diatomic molecules in vacuum gas oil is mainly due to the nature of the crude or residue from which it was obtained. Molecules such as sulphur react more readily than carbon and hydrogen to incorporate oxygen. Oxidation is the primary part, in the context of the ageing process, evaporation or volatilisation and degradation associated with photo-degradation by light. On the other hand, metal atoms such as nickel and vanadium are present only slightly, approximately 1 to 2%. The molecular structures of these fractions are complex, varying in size and type of chemical bond with each source or mixture. There are three basic types of molecules: cyclic, acyclic and aromatic. Acyclic or paraffinic molecules are linear, three-dimensional, chain-like and fatty in nature. Cyclic, or naphthenic, are three-dimensional, saturated carbon rings. Aromatics are flat, stable carbon rings that clump together easily and have a strong odour. All these types of molecules interact to modify the behaviour, individual characteristics and physicochemical specifications of vacuum gas oils [v].

1.6.2. Characterisation

This is done by means of physical properties, structural features and individual analysis of its components. Analyses include: Coking Value (Carbon Residue), Density (API Degrees), Simulated Distillation, Determination of Relative Molecular Mass (Molecular Weight), Determination of S.A.R.A. Fractions, Determination of Carbon (C) and Hydrogen (H), Determination of Nitrogen (N), Determination of Sulphur (S), Determination of Metals. These properties were explained in the section on pitch characterisation with the exception of a new analysis of structural features, called discriminated aromatics analysis (gas chromatography).

1.6.3. Average Molecular Structure

The structural relationship and conformation of vacuum gas oils is very complex and depends directly on the type of crude oil from which they originate. For their elucidation, it must be assumed that they are structures that can be approximated by the results of the various

qualitative and quantitative analyses. On average
They have aliphatic chains of more than 5 carbons, intramolecular sulphur atoms of the thiophene type, a greater number of naphthenic structures than pitches and are more flexible molecules, as well as being precursors of the latter. They have mono-, di-, tri- and tetraaromatic structures, which makes the existence of peri-condensed structures uncommon. Some of these average molecular structures will be presented below (Figures 4 and 5):

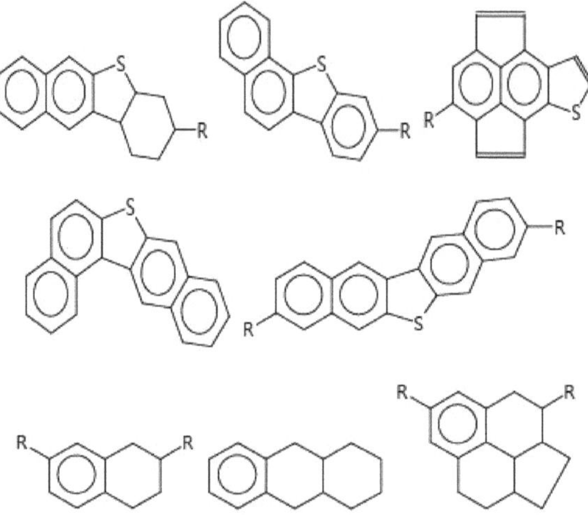

Figure 4. Average molecular structures of saturated, mono-aromatic and polyaromatic polyaromatic GOV [lx].

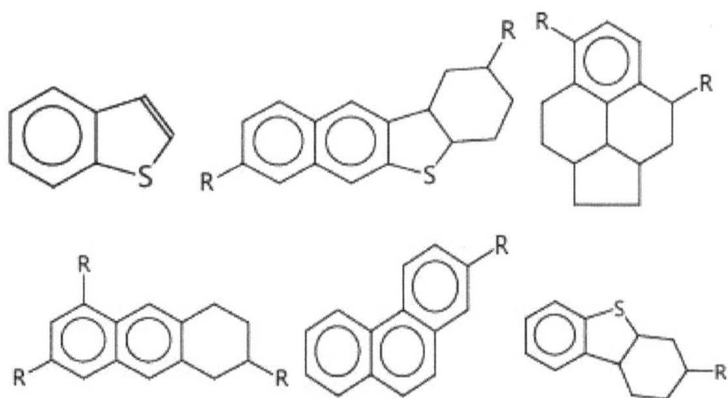

Figure 5. Average polyaromatic molecular structures of GOV [lx].

1.7. Theoretical Foundations for Establishing the Thermodynamic Model for the Conversion of Vacuum Gas Oil to Petroleum Tar Pitch

1.7.1. Gibbs Free Energy Standard Formation Standard and Entropy

For a chemical reaction, the Gibbs free energy is the variation of the free energy G resulting from the conversion of the stoichiometric numbers of moles to products of the pure and separate reactants, each in its standard state T. Thus we have that:

$$\Delta G_T^{\circ} = \sum_i v_i G_{m,T,i}^{*} \quad (13)$$

If the reaction is the formation of a substance from its elements in their reference forms, $\Delta G^{\circ}{}_T$ is the standard Gibbs energy of formation and $\Delta/\ G_T$ of the substance. Since the formation of an element from itself does not involve any transformation, the standard Gibbs energy of formation of the substance will be zero. Thus in the physical sense the variation of ΔG is only applicable to processes with ΔT=0, which leads to:

$$\Delta_f G_T^{\circ} = \sum_i v_i \Delta_f G_{T,i}^{\circ} \quad (14)$$

To obtain the values of Δ/G° *from G=H-TS,* it is established that for an isothermal process ΔG=ΔH-TΔS. Then if the process is for the formation reaction of a substance i, the variation of the Gibbs free energy will be given by:

$$\Delta_f G_{T,i}^{\circ} = \Delta_f H_{T,i}^{\circ} - T\Delta_f S_{T,i}^{\circ} \quad (15)$$

Where:

Δ_{fSTi}= Calculated from the tabulated values of the entropies $S^{\circ}{}_{mT}$ of substance *i* and its elements.

Δ_{fHTi}= Is tabulated

Once these values are known, they are introduced into the equation allowing the variation of the standard Gibbs free energy of formation $\Delta_f G^{\circ}$ [lxi] to be tabulated. Table 6 summarises the trend of the Gibbs free energy as a function of reaction type and orientation:

Table 6. Trend of Gibbs free energy as a function of reaction type and shift of the chemical equilibrium of the reaction.

and displacement of the chemical equilibrium of the reaction.

ΔG	*Type of reaction*	*Shift of chemical equilibrium*
-	Exothermic	Reagents^ Products
0	Balance	Reagents = Products
+	Endothermic	Reagents^ Products

Table 7 shows the trend of the thermodynamic parameters as a function of the spontaneity of the reaction.

Table 7. Trend of thermodynamic parameters as a function of reaction spontaneity.

spontaneity of the reaction.

ΔH	ΔS	ΔG	Result
+	+	+	Non-spontaneous at low temperature
+	+	-	Spontaneous at high temperature
-	-	+	Non-spontaneous at high temperature
-	-	-	Spontaneous at low temperature
+	-	+	Non-spontaneous at any temperature
-	+	-	Spontaneous at any temperature

In discussing these processes we see that:

- In irreversible (spontaneous) processes AG < 0; it represents a tendency towards the final state and in a chemical reaction indicates that the formation of products is favoured. Conversely, a value of AG > 0 represents a spontaneous tendency for reactants to from products.

- In reversible (equilibrium) processes AG = 0; represents the condition of a stable system where the trend between the two states is the same and a chemical reaction indicates that a state has been reached where both reactants and products are in chemical equilibrium.

Since, for a given reaction, it is possible to calculate the value of AH and AS, then the value of AG will be given by the values of these parameters.

1.7.2. Normal Equilibrium Constant

Given a chemical reaction $0 \leftrightarrow \sum_i v_i A_i$ with stoichiometric coefficients v_i, the chemical equilibrium condition set is $\sum_i v_i \mu_{i,eq} = 0$, where $\mu_{i,eq}$ is the equilibrium chemical potential or partial molar Gibbs energy of species i. By choosing a normal state for each species i to obtain a suitable expression for μ_i (chemical state potential of a substance i), the activity ai of i can be defined as:

$$a_i \equiv e^{(\mu' - \mu'^{\circ})/RT} \qquad (16)$$

Where:

a_i= activity dependent on the choice of the normal state

μ_i= Chemical potential of i in the reactive mixture

$\mu°$ =Normal or standard chemical potential

Knowing that the activity a_i of substance in any solution, ideal or otherwise, is given by

$$a_i = \exp[(\mu_i - \mu_i^{\circ})/RT] \qquad (17)$$

Taking the logarithms of equation (17) we have:

$$\mu_i = \mu_i^{\circ} + RT\ln a_i \qquad (18)$$

Substituting the equilibrium condition $\sum_i v_i \mu_{i,eq} = 0$ en (18), we obtain:

$$\sum_i v_i \mu_i^{\circ} + RT \sum_i v_i \ln a_{i,eq} = 0 \qquad (19)$$

Where:

$a_{i,eq}$ = Equilibrium value of activity a_i

$\sum_i v_i \mu_i^{\circ}$ = Increase or variation of the normal Gibbs energy AG∘

Having to:

$$\sum_i v_i \ln a_{i,eq} = \sum_i \ln(a_{i,eq})^{v_i} = \ln \prod_i (a_{i,eq})^{v_i} \qquad (20)$$

Equation (20) becomes:

$$\Delta G^{\circ} + RT \ln \prod_i (a_{i,eq})^{v_i} = 0 \qquad (21)$$

Defining k° equilibrium constant as the product in the latter equation (21), we say that:

$$\Delta G^{\circ} = -RT\ln K^{\circ} \quad (22)$$

$$\Delta G^{\circ} \equiv \sum_i v_i \mu_i^{\circ} \quad (23)$$

$$k^{\circ} \equiv \prod_i (a_{i,eq})^{v_i} \quad (24)$$

From equation (22), the value of the equilibrium situation is only reached when the activities are such that $\prod_i (a_i)^{v_i}$ equals the constant of equilibrium k°.

CHAPTER 2

METHODOLOGY FOR THE PRODUCTION OF BREA

This chapter presents a methodology for obtaining petroleum tar pitch base, the techniques for characterisation and conducting experiments, together with the operation of instruments and equipment.

2.1. Materials and reagents

The separation, conversion and characterisation of the reaction products was carried out in the laboratories of the Technical Management of Residuals and Heavy and Extra Heavy Crudes, as well as in the General Laboratories Management, both located in PDVSA, which have the infrastructure and equipment required for vacuum distillation and thermal cracking of different petroleum fractions. The sample used is an aromatic product produced from PDVSA's technology, whose vacuum distillation bottoms have properties suitable for the production of petroleum tar pitches (BAP). The inert pressurisation gas used during the thermal cracking reactions was nitrogen bottled by BOC Gases de Venezuela, C.A.

2.2. Equipment

The equipment described below was used:

- Distillation equipment made of high-temperature borosilicate glass, HS Matrin Inc. brand, consisting of a balloon with thermowell, heating mantles, an upper condenser to which the vacuum line is attached, a drag separator, a secondary condenser containing a cooling head, a main summary line, a product receiver. Attached to the cooling head section is the vapour sensor (manometer) and temperature sensor (recorder). The parts are connected by vacuum joints for easy maintenance. The vacuum pump, temperature recorder and heating mantle described below are also attached to the distillation unit:
- AEG vacuum pump type AMEB model 71PY4R3N, 0.2 KW, 110 volts, 1700 1/min. Maximum depressurisation value is 1 mmHg.
- Omega temperature recorder model Termotrol 2000, operating in a temperature range (0 - 999) °C with 10 channels, with contacts made of silver. The measuring lines are made of copper.
- Electromantle heating mantle Model EM 2000/c MK1, 500 W, T Max 450 °C 110 Volt, 50-60 Hz.
- Batch type reactor without agitation manufactured at PDVSA, with a capacity of 100 mL, made of type 316 stainless steel, which withstands up to 3000 psig and 500 °C. For the tests carried out, the equipment was fitted with a WHITEY model 12Ek079 300 cc stainless steel 1500 psig cylinder, a Tescom Industrial Controls model 26-1765-25 pressure regulating valve with a maximum capacity of 1500 psig calibrated according to the pressure limit established for the reactor, two Nutro Company model SS-454T quick closing valves and two WHITEY model SS-D55/V54 needle valves.
- Tecam® brand fluidised sand bath furnace, model SBL-2D with Tecam® brand temperature controller, model TC4D.
- Temperature recorder (K-type thermocouple thermometer), brand Omega, Engineering Inc, model 650, 10-channel working temperature range (0 - 999) °C, associated error ± 1°C.
- Gas trap, capacity 500 mL, made of glass (borosilicate).

2.3. Instruments

- Mettler Toledo, model SB32001 Delta Range, (132100) g, associated error ± 1 g. 2. Mettler Toledo, model PB3000-S Delta Range, working range between (0.5- 3100.0) g, associated error ± 0,1 g.
- Manual "U" type mercury manometer, made of glass by Petróleos de Venezuela S.A., working range 0.1 to 5.0 mmHg, reading 0.1 mmHg.
- Type K thermocouple: Consists of two wires of different metals, one positive (+) of iron and one negative (-) of a copper-nickel alloy. These wires are soldered at one end and terminate in a pin. The measuring range is (-210 to 1200) °C.

2.4. Obtaining G.O.V. from Vacuum Distillation of Aromatic Product Commercial Sample

It was carried out at a shear temperature such that the softening point and viscosity conditions for an anode grade petroleum tar pitch were reached. To establish these conditions, distillations were carried out at intervals between 460 °C and 510 °C, this is known as the equivalent atmospheric temperature (TAE) and is the starting point for calculating the Ttope temperature of the system at a given operating pressure. The calculation of (Ttope) was based on the following correlations:

$$A = \frac{6,761559 - 0,987672\log_{10} P}{3000,538 - 43,00\log_{10} P} \quad (26)$$

For a pressure of less than 2 mmHg:

$$A = \frac{5,994295 - 0,972546\log_{10} P}{2663,129 - 95,76\log_{10} P} \quad (27)$$

For a pressure greater than or equal to 2 mmHg:

$$TAE = \frac{748,1A}{[1/(T + 273,1)] + 0,3861A - 0,00051606} - 273,1 \quad (28)$$

where A= dimensionless constant; where:

Where:

P= operating pressure in mmHg.

T= buffer temperature observed at the thermocouple in centigrade (°C).

TAE= equivalent atmospheric temperature in degrees Celsius (°C). The TAE is the temperature at the boiling point of the sample, i.e. at 760 mmHg.

The procedure used to perform the vacuum cuts or distillations according to adaptations of ASTM D-5236-03 and ASTM D-1160-03 is presented below.

- Install the equipment and adjust.
- Verify that the vacuum pump readings are below 5 mmHg.
- Weigh and record the measurement on the balance.
- Transfer the sample to a 2 L distillation still up to 60 % of its capacity.
- Adjust and check the ignition of the heating blankets so that the upper blanket is at a higher temperature than the lower blanket (in the range of 70-60 volts respectively).
- Stop the distillation when it reaches the stop temperature, according to the equivalent atmospheric temperature calculation.
- Extracting petroleum tar pitch and vacuum gas oil distillate.

2.5. Thermal Cracking of G.O.V. at Moderate Severity Conditions in Batch Reactors

The test was carried out in a non-stirred batch reactor manufactured by Petróleos de Venezuela S.A., with a capacity of 100 mL and a fluidised bed heating system with sand. The procedure used to achieve the formation matrix and establish the optimum conditions for oil tar pitch formation is described below:

- Regulate the air flow by progressively decreasing it as the temperature rises until the temperature set in the forming matrix is reached.
- Charge the reactor to 75 % of its capacity.
- Close the reactor by adjusting so that the graphite seal is over the groove in the lid.
- Purge the reactor by injecting nitrogen gas (N_2) and perform the leak test.
- Immerse the reactor in the sand furnace.
- Let the reaction run at the conditions proposed in the formation matrix, cool and open the reactor.

The conditions to which the GOV sample from the vacuum distillation of the commercial sample, whose main characteristic is the high content of cyclic aromatic compounds, was subjected to are shown by means of an experimental matrix.

- .5.1. Construction of a Base Formation Matrix for the Petroleum Tar Pitch

The oil tar pitch formation matrix from GOV obtained from vacuum distillation was based on typical parameters to be taken into consideration when performing thermal cracking reactions. In that sense, variable temperatures and residence times were proposed, keeping the system pressure constant at 250 psig as shown in Table 8. The selected ranges for temperatures, residence times and working pressure were based on operational conditions reported in different patents related to the area (430 °C, 30 min and 250 psig).

Table 8. Temperature Vs. residence time.

Temperature (°C)	Time 1 (min)	Time 2 (min)	Time 3 (min)
420	60	45	35
430	60	45	35
440	60	45	35
450	60	45	35

2.6. Obtaining BAP from Vacuum Distillation of Petroleum Tar Pitch Base

The petroleum tar pitch base was treated following the procedure described in section 2.5. Vacuum Distillation of Aromatic Product Commercial Sample, in order to obtain petroleum tar pitch from thermally cracked vacuum gas oil.

2.7. Characterisation of Aromatic Products

According to the type of properties to be evaluated, the characterisation of materials can be subdivided into three groups: physical properties, structural features and individual component analysis.

2.7.1. Physical Properties

It was carried out by means of analytical methods standardised by ASTM (American Standard and Testing Materials) and its own analytical methods developed by PDVSA's General Laboratories Management.

2.7.1.1. Softening Point

It was carried out on the Herzog model HRB 754 digital ring-and-ball softening point measuring device. This arbitrary property was measured when two horizontal discs of sample, immersed in bronze rings, were heated at a controlled rate in a liquid bath while a steel pellet was attached to each disc. The softening point was reported as the average of the temperatures at which the two discs softened sufficiently to allow each ball, enveloped in the asphalt, to fall to a distance of 25 mm (1.0 in). The procedure employed was as follows: In a beaker, used as a bath, glycerine was placed, so that the temperature of the sample would reach the desired boiling range (25-30 °C), subsequently, the sample was heated until fluidity was achieved, then, it was introduced into the metal rings on aluminium foil, taking care that it was well distributed in the ring, so that the surface of the sample coincided with that of the upper face of the rings. According to the ASTM D 36-06 standard, the bath and the sample must reach a temperature of 30 °C to start the test, then one ball of the equipment was placed on each ring with sample and immersed in the glycerine bath, then the ignition of the equipment was activated. By the effect of gravity the balls were immersed to the bottom of the beaker, which are detected by the laser of the equipment, and automatically the equipment indicated the drop temperature, which is known as the softening point.

2.7.1.2. Density

The relative density was determined in a Pyrex pycnometer with a capacity of 10 mL. According to ASTM D 70-03. The density of the sample was calculated from its mass and the mass of water that was displaced by the sample at the time of filling the pycnometer using the following equation:

$$\text{Relative Density} = \frac{(C-A)}{[(B-A)-(D-C)]} \quad (29)$$

Where

A = mass of the pycnometer.

B = mass of the water-filled pycnometer.

C = mass of the pycnometer partially filled with sample.

D = mass of pycnometer + sample + water.

Density of the sample = Relative density x Density of water

Density of water @ 15 °C = 997.0 Kg/m^3

The methodology used was as follows:

- Weigh the pycnometer (clean and dry) to the nearest 1 mg (A).
- Fill the pycnometer with distilled water and place in a bath for a period of 30 minutes. Then remove the pycnometer, dry its contour and weigh to the nearest 1 mg (B).
- Fluidise and shake the sample (at more than 55 °C above its softening point, for a time not exceeding 60 min), add a portion to the pycnometer until about % of its capacity is filled, without the sample touching the walls surrounding the free volume. Allow the temperature of the pycnometer and its contents to cool to room temperature for a period of 40 min, and then weigh to the nearest 1 mg (C).

- Fill the free volume of the pycnometer with water without bubbles. Press the stopper firmly. Place the pycnometer with the sample and water in the bath for not less than 30 min. Dry and weigh to the nearest 1 mg (D).

The specific density or API gravity was determined using an ERTCO (18-23)°21 H glass hydrometer and a thermometer. Following the ASTM D 1298-99 standard. The property was measured using the following scheme of work, the hydrometer and the thermometer were placed at 5 °C inside a cylinder, the sample was added avoiding the formation of bubbles and the measurement was taken in a place where the temperature did not vary by more than 2 °C. Once these conditions were reached, the density and the temperature at which the sample was found were recorded.

2.7.1.3. Viscosity

Viscosity is measured by rotation of a rod attached to a geometric element (cylinder shape) immersed in the sample liquid. The spindle rotates at a known rate and the instrument reports the torque required for the cylinder to rotate. Taking into consideration the rotational speed, the torque measurement and the cylinder characteristics, the instrument calculates the velocity of the fluid of interest. The procedure according to ASTM D 4402(06) was based on turning on the HT-104 temperature controller, then the temperature reading was changed from degrees Fahrenheit to degrees Celsius, then the desired temperature condition was set, the thermosel was started to heat up, the spindle was placed and immersed without being completely covered by the sample. Finally, record the measurement in units of Pascal/Second or centipoise.

2.7.1.4. Coal residue

It was carried out in a Tanaka furnace, model ACR - M3, using nitrogen gas (N_2), according to ASTM D 4530-06. This property was measured by placing a quantity of sample of known weight (0.15 g), in a glass vial and heated to 500°C under an inert atmosphere, in a controlled manner and for a specified time. Due to the effect of decreasing the partial pressure of the hydrocarbons and the entrainment effect caused by the nitrogen, during the severe heating, the volatiles produced during the coking reactions are released leaving only the carbonaceous type residues in the container, which are indicated as "% Carbon Residue (Micro Method)" of the original sample. When the expected result is below 0,10 % by weight, the sample may be distilled to produce 10 % by volume of background before testing.

2.7.1.5. Solubility

It was carried out following a procedure similar to an internal PDVSA standard using a centrifuge and an explosion-proof cooker.

The methodology used consisted of heating and homogenising the aromatic sample and weighing it (approximately 5 g) in a centrifuge tube containing 50 mL of toluene, gauging it to 100 mL in order to dissolve the sample. Once dissolved, the sample was centrifuged for 20 min at 1500 rpm and the precipitate was decanted. All the insoluble material was washed with toluene and then the tube was rooted up to 100 mL, repeating the procedure until the washes were clear. Finally, the precipitate was dried in the oven at 105 °C and weighed to calculate the mass percentage of toluene insolubles in the samples using the following ratio:

$$I.T = \frac{(B-A)}{C} * 100\% \qquad (30)$$

Where:

A= Weight of dry, clean centrifuge tube in grams.

B= Weight of dry insoluble substances in grams plus the weight of the dry, clean centrifuge tube in grams.
C= Weight of the sample in grams.

2.7.2. Structural Features

2.7.2.1. Simulated Distillation

It was carried out using an Agilent Technologies gas chromatograph, model 6890N with an Agilent Technologies injector, model 7683 B Series. The boiling point distribution range of the sample was obtained by simulated distillation through a gas chromatograph. A packing or capillary tube was used to elute the hydrocarbon components of the sample in increasing order of their boiling points, the column temperature was described or increased at a linear rate and the area under the chromatogram was recorded throughout the analysis. The boiling points were assigned at the same time as the axis of a calibration curve, obtained at the same chromatographic conditions by analysis of a mixture of known hydrocarbons spanning the range of expected boiling points in the sample. From these data, the distribution range of an unknown sample could be established. The standards used were:

ASTM D-2887. Flame ignition ratios are: hydrogen 40 mL/min, air 450 mL/min, MK up 40 mL/min. Drag gas velocity (Helium) 30 mL/min. Detector temperature 360 °C. Injector temperature 350 °C. Sample 0,2 g in 10 mL carbon sulphide and inject 0,2 µE.

ASTM D-7169. Flame ignition ratios are: hydrogen 40 mL/min, air 450 mL/min, MK up 25 mL/min. Drag gas velocity (hydrogen) 15 mL/min. Detector temperature 435 °C. Injector temperature ramps up with the oven at a rate of 15 °C/ min from 50 °C to 425 °C. Sample 0,2 g in 10 mL carbon sulphide and inject 0,2 µE.

2.7.2.2. S.A.R.A

It was analysed using an IATROSCAN equipment, model MK-6s, which combines thin layer chromatography with a flame ionisation detector on quartz rods coated with silica gel. The surface of the silica gel and alumina, being covered with fixed dipole sites, induces the dispersion of the electrons of the organic compounds, so the separation of the different fractions occurs in the following way:

- Saturated: the portion of the sample elutes from the column with n-heptane.
- Aromatics: the fraction of the sample that does not elute with n-heptane, but is eluted with a solvent of moderate polarity.
- Resins: the fraction that elutes with a solvent of higher polarity.
- Asphaltenes: due to its characteristics, in this fraction part of the sample precipitates in solution with n-heptane, under certain specific conditions. The solvents used for each of the fractions were: saturated n-heptane, Aromatics toluene, Asphaltenes mixture of n-heptane and isopropanol in a ratio of 5:95, FID detector flow rate Air 2000 mL/min hydrogen 160 mL/sec. Reading speed 30s/scan sample quantity 2426 mg in 1 mL of a 1:1 chloroform-toluene mixture, according to PDVSA internal standard.

2.7.2.3. Vapour Pressure Osmometry (V.P.O.)

This property was measured by means of the vapour pressure of the system, part of the sample was dissolved with an appropriate solvent, this quantity was recorded. A drop of this solution and a drop of solvent were suspended side by side, separated by a thermal curtain in a closed chamber saturated with solvent vapour. Since the vapour pressure of the solution is lower than that of the solvent, the solvent condensed on the sample and caused a temperature drop between the two drops. This resulting temperature change was measured and used to determine the relative molecular mass (molecular weight) of the sample with reference to a previously prepared calibration curve. The methodology used is derived from ASTM D-2503 and is described as follows: the solvent used was selected and the osmometer cup was filled with it, the 25 mL volumetric flask was weighed 0.3 to 0.6 g and diluted with solvent to capacity, the syringes were filled with sample and solvent, the sample was added to the thermal curtain and the experiment was allowed to run. Running a series of samples the temperature values were recorded, then the calibration curve was used to obtain the molarity by installing the vapour wick and filling the slots inside the wick, the sample cup was placed and the values were recorded.

2.7.2.4. Discriminate Aromatics Analysis

It was carried out in an HP gas chromatograph, model 5890 series II, this equipment has an internal furnace that reaches a maximum temperature of 400 °C with a flame ionisation detector of 350 °C, the capillary injector with flow division (split), the flow ratio in split mode is 133:1. This equipment has an HP mass selective detector, model 5970 series, with a flow rate of 100 mL/min and consisting of a 70 eV electron impact ionisation source, a quadrupole mass analyser and an electromultiplier detector. The mass detection range is between 30 and 700 Dalton. The sample is injected at a volume of 1 μL. The analysis methodology is for the exclusive use of Petróleos de Venezuela, S.A.

2.7.3. Elemental Analysis

2.7.3.1. Carbon and Hydrogen

It was carried out according to ASTM D 5291-02 using a furnace with infrared detector Leco, Model CHNS-932, the measurement was made by oxidising the sample inside a crucible at high temperatures, then the calcined sample and its gases passed through the I.R. detector, which gave the percentages of carbon and hydrogen present in the sample. The methodology used is explained as follows: a sample amount of 2 to 5 mg was placed in a tin crucible and oxidised at 950 °C, a secondary furnace connected transferred the sample to the I.R. detector, which is calibrated on the basis of carbon dioxide and water standards.

2.7.3.2. Sulphur

This analysis was performed using an Axios Petro Panalytical X-ray fluorescence spectrometer. The property was taken for the identification and quantification of the chemical elements by establishing a relationship between the X-ray emission and the nuclear charge of the atoms, as the sample was placed in the X-ray beam and the peak (Ka sulphur line intensity at 5.373 Â) was measured. The intensity, which was measured at a recommended wavelength of 5.190 Â (5.437 Â for a tubular Rhx target) was subtracted from the peak intensity. The resulting net count rate was compared with the calibration curve that was previously prepared to obtain the sulphur concentration in mass percentage.

The procedure used was as follows: the sample is placed in a cell at a minimum of three quarters of the cell capacity. The sample was introduced into the X-ray beam allowing the optical path to reach equilibrium. The intensity of the radiation at Ka 5.373 Â sulphur was determined by counting the precise angular rate measurements for this wavelength configuration. The corrected rate was determined and the sample concentration quantified.

2.7.3.3. Nickel and Vanadium

It was analysed according to PDVSA procedures. The equipment used was an optical emission spectrometer with inductively coupled plasma Varian Vista Pro CCD Simultaneous

ICP OES, which made it possible to obtain the signal of the presence of the element, and the value in concentration of the element proportional to the intensity of the light detected (Quantitative Analysis - Qualitative Analysis). The principle of this technique is the detection and quantification of light emitted by an atom or molecule that has undergone a previous excitation process in a gas at high temperature (plasma). When a set of atoms is subjected to very high temperatures, a large number of atoms are excited and subsequently emit light when they return from an excited state to a lower energy state, the intensity of the light emitted being proportional to the population of atoms in the excited state.

2.7.3.4. Nitrogen

It was analysed according to ASTM D5762. Using an ANTEK 9006 UV-VIS absorption/emission detector. The measurement was performed by placing the hydrocarbon sample on a cell at room temperature. The cell with the sample is brought to high temperature for combustion inside the tube where the nitrogen is oxidised to nitric oxide (NO) in an oxygen atmosphere. NO in contact with ozone is converted to nitrogen dioxide (NO_2). The light emitted by excited NO2 decays and is detected by a photomultiplier tube and the resulting signal is a measure of the nitrogen contained in the sample. The sample preparation methodology was carried out in accordance with practice D 4057, diluting 0.05 g in 1.6 g xylene in a 2 mL vial setting a range of 10 to 100 ppm.

2.8. Computational Analysis for Thermodynamic Modeling

2.8.1. AQC® Kinetic Model Approach - VB.V.1.0

For the use of this method of molecular structuring of each of the S.A.R.A. fractions, a bibliographic review was carried out to establish the basic formats for the tentative structures for each group. Then, based on experimental data, elemental analysis and SARA analysis of the charge, it was possible to perform a mass balance closure that allowed the consolidation of the proposed structures.

Finally, the boiling points of the theoretical representative molecules of the feed were calculated according to the Meissner method. The validation of the proposed molecules was carried out by comparing them with the simulated distillation curve of the initial charge. The coincidence of the boiling points of the proposed molecules and the initial charge is indicative of the adjustment of the proposed molecules to reality, since the boiling point is directly related to the molecular structure.

2.8.2. Balance of the General Equation of BAP Formation

This balance was carried out in order to obtain a generalised molecule for both GOV and BAP, which in turn comes from the sum of each of the representative molecules that make up both GOV and BAP, which contributes significantly to simplifying the thermodynamic model.

2.8.3. Program for the Calculation of Thermochemical Properties. QBTherm (TM) V3.0 (Copyright 1992-2024)

This program is based on the Group Addition Method of Perry's HandBook 1984. During the development of this study, it was used to calculate the values of Cp, enthalpy of formation and entropy of formation for ideal gases of organic compounds. In order to use it, the different functional groups that make up the compound must be identified, as well as the quantity of these functional groups. With this information, the properties were calculated using the thermog2/fortran program, which is included in the QBTherm (TM) V3.0 program.

CHAPTER 3

RESULTS AND DISCUSSIONS ON THE PITCH PRODUCTION

This chapter presents the results obtained from the experiments carried out following the sequence described in the experimental methodology.

3.1. Obtaining GOV from Vacuum Distillation of Aromatic Product Commercial Sample

Table 9 shows the percentage yields of vacuum gas oil, petroleum tar pitch from vacuum distillation of the aromatic starting product.

Table 9. Yields of petroleum tar pitch and vacuum gas oil from commercial sample at 490 °C cut-off temperature.

from commercial sample at 490 °C cut-off temperature.

Experience	Cutting temperature (°C)	Performance Pitch (%)	Performance GOV (%)	Loss (%)
A	490	46,99	47,44	5,57
B	490	51,96	42,75	5,29
C	490	49,00	46,03	4,97
Average	490	49,32	45,41	5,27

In order to establish the cut-off point for obtaining a petroleum tar pitch suitable for use as a binder, vacuum distillations were carried out at intervals between 460 °C and 510 °C, evaluating the softening and viscosity points as inspection properties to preliminarily define the quality of the product. The equivalent atmospheric temperature (EAT) of the cut-off was 490 °C. As described in the previous chapter, the aromatic sample was physically separated under vacuum conditions to minimise its thermal degradation.

The results obtained (Table 9) are consistent with each other and fall within the expected range of yields that can be predicted from the distillation curve of the starting aromatic product. The differences observed in Table 9 can be attributed to initial charge segregation and slight pressure and temperature fluctuations observed during the experiment, as well as to the properties of the charge used. On the other hand, the losses reported in experiments A, B and C are attributed to the loss of material on the walls of the balloon and the distillation equipment, as well as to the escape of light product, since although the cooling system is efficient, when an increase in pressure is evident in the system due to the high formation of vapours, a fraction of the GOV in vapour phase passes directly to the suction pump, as a consequence of a low contact time between the vapours and the cold surface.

3.2. Study of the Thermal Reactivity of G.O.V. at Moderate Severity Conditions in Batch Reactors

In this section, thermal reactivity is studied through the effect of thermal cracking at moderate severity conditions (temperature and pressure) in a set reaction time on vacuum gas oil, in order to achieve the generation of petroleum tar pitch base. The thermal cracking reaction of GOV was carried out in triplicate in Batch reactors and an inert atmosphere of 250 psig N_2, following the formation matrix described in section

2.5.1. The average yields of petroleum tar pitch base with respect to the set temperatures and reaction time are shown in Table 10. These conditions were selected, as they are the variables that affect the quality and yield of the heat conversion products.

Table 10. Average yield of petroleum tar pitch base (BBAP) versus temperature and reaction time.

(BBAP) with respect to temperature and reaction time.

Temperature (°C)	BBAP yield (%) at t= 60 (min)	BBAP yield (%) at t= 45 (min)	BBAP yield (%) at t= 35 (min)
420	82,14	82,90	83,83
430	73,67	76,04	81,00
440	60,57	64,30	71,70

Knowing that in the thermal conversion process very complex reactions occur such as: dealkylation (loss of aliphatic side chains, formation of paraffins, olefins and dealkylated naphthenes with or without substitutions); partial decomposition of functional groups for the formation of other molecules; alkylation; cyclisation; isomerisation and dehydrogenation of naphthenes and cyclo-olefins, to generate aromatics among many other molecules. These reactions are responsible for generating lighter products, as well as producing heavier residues with a higher carbon-to-hydrogen ratio (C/H) than the feed, limiting the levels of thermal cracking due to the stability of the product generated. Generally speaking, the amount of high molecular weight chemical compounds formed increases at higher temperatures. Nevertheless, it can be affirmed that the results obtained guarantee a production of at least 60% of petroleum tar base from GOV, a product that is certainly more polycondensed than the starting GOV. Likewise, when comparing these obtained percentages, it is observed that the tendency of the oil tar pitch base production yield is to decrease with increasing temperature and residence time, due to the fact that at these conditions the severity of thermal cracking is greater, so there is a greater release of volatile material attributed to the light fractions or aliphatic chains, leaving inside the reactor the fraction that has the most condensed structures. Therefore, it could be seen that with increasing temperature and residence time, gas production increases, which is an indication that the thermal cracking reactions are favoured, as well as the production of petroleum tar.

The properties that were selected for the evaluation of the oil tar pitch base were micro carbon residue (M.C.R.) and toluene insoluble (T.I.) because they are considered fundamental in determining the applicability of oil tar pitch as a binder in electrode production. The rationale behind the analyses on the oil tar pitch base samples and not on the pitch is the direct correlation between their properties. Thus, a high RMC and I.T content of BBAP immediately translates into similar properties of the pitch that would be produced.

Table 11 presents the physical properties (RMC) and (I.T) of the petroleum tar pitch base at defined temperature and time. For the reporting and analysis of the results, the average values of the tests performed in triplicate were used. The GOV samples treated at 420 °C did not show any appreciable chemical reactions in any of the cases of study, as no mass loss attributable to vapours (below 5%) was observed and the physical appearance of the sample did not change. Also, at 450 °C conditions, the presence of solid phase inside the reactor was observed and this condition is discarded, since the material does not present suitable properties as a binder. Among the experimental errors, we can mention those associated with the measurement and weighing of the load used in the experiments, as well as the small losses due to mechanical drag-out during the development of the reaction.

Table 11. Physical properties of petroleum tar pitch base (BBAP).

Temp. (°C)	Time 60 (min)		Time 45 (min)		Time 35 (min)	
	RMC (%P/P)	I.T (%P/P)	RMC (%P/P)	I.T (%P/P)	RMC (%P/P)	I.T (%P/P)
420	*S/C	*S/C	*S/C	*S/C	*S/C	*S/C
430	17,0	2,32	15,4	1,16	7,37	0,51
440	22,5	9,78	19,5	5,50	14,0	2,63
450	**F/C	**F/C	**F/C	**F/C	**F/C	**F/C

*S/C= No physical change was observed in the sample.

**F/C= Formation of solid agglomerated material was observed.

From Table 10 and 11 it can be said that the results of micro carbon residue and toluene insolubles reported at temperature conditions of 440 °C and residence times between 45 and 60 minutes are the most suitable, because from these conditions appropriate RMC and I.T. are obtained to obtain petroleum tar pitch (BAP), once the vacuum distillation of the petroleum tar pitch base (BBAP) is carried out. The relationship between BBAP and pitch can be established by means of the following equations:

$$\%\ \text{BAP yield} = \frac{RMC(BBAP)}{RMC(BAP)} * 100\% \quad (30)$$

$$\%\ \text{BAP yield} = \frac{I.T(BBAP)}{I.T(BAP)} * 100\% \quad (31)$$

Figure 6 shows that when the system is subjected to high residence times there is greater formation of micro-coke residue or heavy components, which is closely linked to the API gravity, the asphaltic content of the product, the aromaticity of the pitch and the properties (density and viscosity) that influence the binding capacity of the pitches, this tendency is limited by the formation of experimental coke. In the same vein, Figure 7 shows the trend of the increase in the percentage of toluene insolubles, which is directly proportional to the residence time of the filler. This effect can be associated with the formation of particles resulting from the break-up of larger molecules and the combination and/or disproportionation of radicals that generate stable products, given in the propagation and termination stage of thermal cracking.

This increase contributes to significantly improve the viscosity of the petroleum tar pitch base, as the average size of the types of compounds attributed to the toluene insoluble (T.I.) have the ability to penetrate into the larger particles of the base, keeping all molecular particles bound together at the intramolecular level. Table 12 shows the microcarbon residue (RMC) and toluene insoluble (T.I.) of GOV and BBA for 440 °C conditions at 45 and 60 minutes.

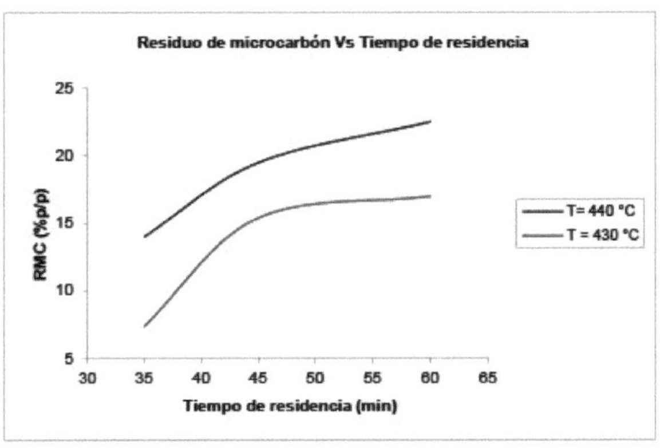

Figure 6. Microchar residue Vs. residence time.

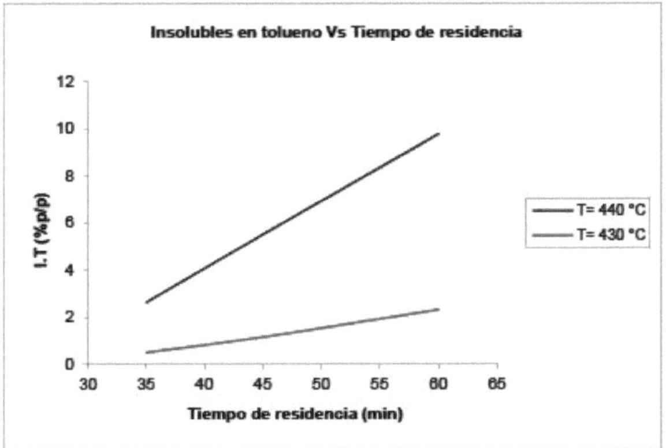

Figure 7. Insoluble in toluene Vs. residence time.

As can be seen, the properties of the petroleum tar pitch base from the higher severity reactions show in general greater changes with respect to the sample treated at 430 °C. This behaviour is logical and expected, since from lower molecular size and weight molecules a higher molecular size and weight fraction can be generated. This behaviour is logical and expected, since from molecules of smaller size and molecular weight a fraction of molecules of larger size and molecular weight can be generated, which is why it can be stated that the higher the severity, the higher the charge conversion.

Table 12. RMC and I.T of the GOV and the average of the reported RMC and I.T results of the reported petroleum tar pitch base (BBAP).

Sample	RMC (% W/W)	I.T (% W/W)
GOV	0,28	*N/A
BBAP at 440 °C at 45 min	19,50	5,50
BBAP at 440 °C at 60 min	22,50	9,78

*N/A= Not applicable for distillate samples.

The large variation of the results obtained between the GOV and all physical properties presented in Table 12 indicates that, in the thermal cracking performed, the number of reactions involved must have been high. According to references, this is possible because the working temperature is highly effective in generating bond breaks of a large number of molecules since their bond dissociation energies are relatively moderate. The observed high rate of thermal cracking occurred despite the fact that the structural chemical reactivity of the families summarised in the bond breaking and non-bond breaking was in the following order:

Paraffins > naphthalenes > olefins > aromatics.

3.3. Obtaining BAP from Vacuum Distillation of Petroleum Tar Pitch Base obtained in the reactor by batching

Table 13 shows the duplicate yields of oil tar pitch and vacuum gas oil from the base load of oil tar pitch from GOV thermal cracking.

Table 13. Performance of petroleum tar pitch and vacuum gas oil from thermally cracked petroleum tar base.

from thermally cracked petroleum tar base.

Experiencia	Temp. cutting (°C)	Rendimiento BAP (%)	Rendimiento BAP (%) GOV (%)	Lost from Material (%)
A	490	40,2	54,7	5,1
B	490	39,9	55,2	4,9
A+B/2		40,1	54,9	5,0

This physical separation was carried out at an equivalent atmospheric temperature (TAE) of 490 °C. Since this procedure was performed using the vacuum distillation method and equipment for obtaining GOV from the aromatic product, errors are assumed to be similar to those previously presented in section 3.1.

From the results reported in this vacuum distillation, it is observed that the percentage yield is typical of those reported for a load whose characteristics are similar, in addition, these represent an improvement to the process of obtaining petroleum tar pitch since, generally, at this cutting condition, yields of around 30 % were achieved. This is due to the fact that the micro-coal and toluene insoluble residue percentages are higher than those generally used to obtain commercial pitches.

3.4. Characterisation of Aromatic Products

The study and evaluation of the properties of the aromatic products; vacuum gas oil (GOV) and petroleum tar pitch (BAP), was carried out through the use of experiments that are given by the physical properties, structural features and individual analysis of the components of the samples.

3.4.1. Physical Properties

Table 14 presents the characterisation of physical properties such as: softening point (P.A), micro carbon residue (MCR), solubility (I.T), density and viscosity, evaluated for both GOV and BAP, to evaluate the evolution of the thermal cracking process.

A comparison of the results in Table 14 shows a significant increase of the described properties as a consequence of the charge conversion. Overall, there is an increasing trend from reactant to product properties, demonstrating the feasibility of producing highly condensed aromatic products from GOV.

There is an increase in all these properties with respect to GOV due to the effect of the working temperature and pressure, 440 °C and 250 psig respectively. As pointed out in previous sections, this marked increase is due to the fact that the activation energy for carbon-carbon, carbon-hydrogen and carbon-sulphur bond breaking, among others, is low, so that at moderate severity conditions a large number of reactions of this type occur.

Table 14. Characterisation of the physical properties of GOV and BAP.

Test	GOV	BAP
P.A (°C)	*N/A	128,000
RMC (% w/w)	0,280	56,000
I.T (% w/w)	*N/A	19,260
Density (g/mL)	1,064	1,215
Viscosity @180 (cP)	2,480	3560,000

*N/A= Not applicable.

Table 15 shows the characterisation of the physical properties of petroleum tar pitch (PTP) and some typical values for commercial petroleum tar pitch (CPTP). Comparing the physical properties of BAP with BAPC, it can be seen that the values are mostly above the typical specification for these products.

Properties such as product viscosity can be appreciably improved by lowering the distillation temperature at which the tar pitch was obtained, which would allow a slight increase in product yield. These characteristics are required so that the baked anode block has the lowest electrical resistivity and the carbonised pitch has a balanced reactivity to air and carbon dioxide with that of coke, in order to avoid localised reactions during operation in the cell.

Table 15. Characterisation of the physical properties of BAP and some typical values for BAPC.

values for BAPC.

Test	BAP	BAPC
P.A (°C)	128	128,5
RMC (% w/w)	56	53,4
I.T (% w/w)	19,26	9,8
Density (g/ml)	1,215	1,21
Viscosity @180 (cP)	3560	< 2000

3.4.2. Structural Features

The structural characterisation of vacuum gas oil and petroleum tar pitch was performed by means of specialised analytical techniques for reagent and product such as: simulated distillation, vapour pressure osmometry (V.P.O.), determination of S.A.R.A. fractions and discriminated aromatics analysis (D.A.A.). The latter applies only to light and middle distillate samples, as the sample must be liquid for the analysis to be performed. In order to comparatively analyse the increase in the light and heavy fractions of the hydrocarbons under study due to the effect of conversion, the simulated distillation curves corresponding to GOV and BAP were plotted.

Figure 8 shows results that demonstrate the trend of the conversion to produce a higher amount of non-evaporable product and the improvement of the reagent properties leading to reactions that lead to the formation of heavier fractions that make up the petroleum tar pitch, as reflected in the vapour pressure osmometry measurements as shown in Table 16.

Figura 8. Simulated distillation of GOV and BAP.

Table 16. Characterisation by vapour pressure osmometry (V.P.O.) of GOV and BAP.

GOV and BAP.

Sample	O.P.V (g/mol)
GOV	239
BAP	593

Likewise, it can be observed that there is an increase in the boiling points of the product with respect to the reagent, this can be attributed to the thermal conversion of the charge, since the higher the boiling point, the greater the complexity of the associated molecules.

From Table 16 it can be inferred that there is a significant increase in the apparent molecular weight of the product with respect to the load, which is attributed to thermal conversion indicating that typical condensation and polymerisation reactions responsible for this increase occurred during the heat treatment. Figure 9 shows the average results for the duplicate tests for the percentages of Saturates, Aromatics, Resins and Asphaltenes (SARA) for vacuum gas oil and petroleum tar pitch.

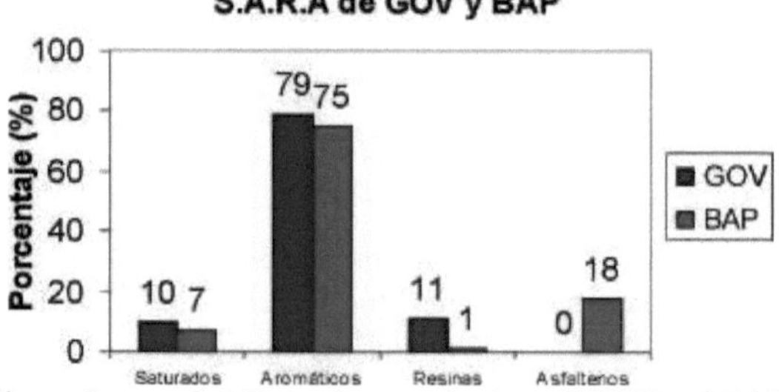

Figura 9. S.A.R.A. characterisation of GOV and BAP.

It is found that due to the thermal effect there is a significant variation in the percentages of saturates, asphaltenes, resins and aromatics, which is consistent with the high variation of physical parameters (density and viscosity) of both the filler and the product. On the other hand, the decrease in aromatics content and the increase in asphaltenes are attributed to the condensation and polymerisation reactions that accompany the thermal cracking of GOV.

Table 17 presents the discriminated aromatics analysis (D.A.A.) performed for GOV, which is intended to give a representation of the types of saturated and aromatic molecules present in the payload. From Table 17 it can be inferred that the main types of polyaromatics present in the cargo are constituted by structural systems ranging from one to four aromatic nuclei, which are flat, rigid and stable molecules that cluster together very easily.

Table 17. Characterisation by analysis of discriminated aromatics for GOV and BAP.

Molecule type	Volume (%)	Weight (%)
Saturated	9,0	9,0
Mono aromatics	13,6	11,7
Di aromatics	12,2	11,3
Tri aromatics	13,2	12,8
Tetra aromatics	26,3	27,1
Penta aromatics	3,6	4,2
Thiophene aromatics	20,4	21,8

Therefore, its chemical reactivity to thermal cracking processes demonstrates the feasibility of producing condensed polyaromatic compounds similar to those contained in this pitch generated from the analysed GOV.

3.4.3. Elemental Analysis

In general, elemental analyses were carried out to determine the percentages of the following elements: Carbon, hydrogen, sulphur, nickel, vanadium and nitrogen in order to relate them directly to the molecular structures that were proposed.

Table 18 shows the elemental characterisation of GOV and BAP. Comparing the results in Table 18 and knowing the highly aromatic nature of GOV and BAP, the higher the distillation temperature, the higher the concentration of polyaromatic species, whose carbon-hydrogen ratio tends to be higher than that of the starting product.

Similarly, the fact that the percentage of sulphur and nitrogen does not vary significantly can be attributed to the fact that these elements are mainly found within the aromatic structures and not in the aliphatic ones.

For this reason, even though the filler was subjected to thermal cracking there will be no significant breaking of heteroatomic bonds, since aromatic compounds possess higher bond dissociation energy than aliphatic compounds, due to contributions from the inductive and resonance effect, causing most of the sulphur and nitrogen to tend to remain in the pitch.

On the other hand, metals such as vanadium, although associated with particles that are dispersed in the starting product (GOV), reflect an increase in vanadium in the pitch as a consequence of the concentration of solids in this fraction.

Table 18. Elemental characterisation of GOV and BAP.

Element	GOV	BAP
C (% p/p)	88,36	89,47
H (%)	7,75	5,24
S (ppm)	2,60	2,50
N (ppm)	0,70	0,70
Na (ppm)	<5,00	<5,00
Ni (ppm)	<5,00	<5,00

V (ppm)	<5,00	18,00

3.5. Computational Analysis for Thermodynamic Modeling

3.5.1. AQC® Kinetic Model Method - VB.V.1.0

Figura 1. GOV and BAP Molecular Structure Approach

After the literature review and based on the characterisation carried out, the representative molecules of the SARA fractions for both GOV and BAP were proposed. For GOV, two molecules for saturates, five molecules for aromatics and three for resins were outlined. For this product, no molecules of the asphaltene fraction were proposed, as this filler does not have them. Similarly, two molecular structures of saturates were proposed for BAP, five for aromatics, one for resins, because the thermal conversion of the charge tends to the low formation of this group and four for asphaltenes, which together totals 12 molecules for BAP and 10 for GOV.

The purpose of proposing several molecules to represent the charge and product of the BAP formation process from GOV was to create a sufficiently flexible thermodynamic model, i.e. the molecules proposed in this study are able to represent different types of feeds and intermediate streams of the process.

Once the representative molecules had been designed, the mass balance was closed, highlighting that the reproducibility of the elemental and SARA analyses of the charge was very precise and satisfactory. As can be seen in Table 19 of the elemental analyses generated for the theoretical GOV molecules proposed.

Table 19. Comparison between the elemental analyses generated by the theoretical molecules and the

Theoretical molecules and experimental analyses for GOV.

Atom	Theoretical (% w/w)	Experimental (% w/w)	Difference	Difference <=5 (%)
C	88,36	88,8	0,4	0
H	7,75	7,8	0,0	0
S	2,6	2,6	0,0	0
N	0,7	0,9	0,2	0

Table 20 also shows the elemental analyses generated by the theoretical molecules proposed for BAP and their error percentages in relation to the experimental analyses.

The comparison between the SARA analyses generated by the proposed theoretical molecules and the experimental GOV and BAP analyses, as well as the difference between each of these and the experimental ones, is summarised in Tables 21 and 22.

Table 20. Comparison between the elemental analyses generated by the theoretical molecules and the

Theoretical molecules and experimental analyses for BAP.

Atom	Theoretical (% w/w)	Experimental (% w/w)	Difference	Difference <=5 (%)
C	89,47	91,60	2,10	2,00
H	5,24	5,40	0,10	2,00

S	2,30	2,30	0,00	0,00
N	0,70	0,70	0,00	0,00

Table 21. Comparison between the SARA analyses generated by the theoretical molecules and the experimental

and the experimental analyses for GOV.

Groups SARA	Theoretical (% w/w)	Experimental (% w/w)	Difference	Difference <=5 (%)
Saturated	10	10	0	0
Aromatics	79	79	0	0
Resins	11	11	0	0
Asphaltenes	0	0	0	0

From Table 20 it can be seen that the difference between the percentage of the theoretical analysis and the experimental value does not exceed 2 percent error, so it was within the expected range (less than or equal to 5 %).

Table 22. Comparison between the SARA analyses generated by the theoretical molecules and the experimental

and the experimental analyses for BAP.

Groups SARA	Theoretical (% w/w)	Experimental (% w/w)	Difference	Difference <=5 (%)
Saturated	7	7	0	0
Aromatics	74	74	0	0
Resins	1	1	0	0
Asphaltenes	18	18	0	0

As in the previous cases, the data in Table 21 and 22 allow to verify by means of the difference between each of the percentages the effectiveness of the method used for the approach of the molecules, as well as the reproducibility of the experimental SARA analyses by the theoretical molecules proposed to represent vacuum gas oil and petroleum tar pitch.

Figura 2. Boiling point assessment

Once the boiling points of the theoretical molecules outlined through the Meissner method [lxii] were obtained, which have codes such as SM1, ARM1, RM1 and ASM1, meaning saturated molecule number 1, aromatics molecule 1, resins molecule 1 and asphaltenes molecule 1 respectively, it was possible to obtain the simulated distillation curve and to observe how much, in %OFF, remains in the different cuts of the distillation, which can be seen in Table 23 where the percentages of evaporation of the experimental simulated distillation (P.E.E), the theoretical evaporation percentage (P.E.T) and the theoretical and experimental boiling points, in addition to the differences between these results, here, the theoretical molecules reproduce very well the data obtained from the experimental simulated distillation of GOV and BAP, the percentage error greatly exceeds the expectations of the established method , since it is below 0.5 %, demonstrating that the selected molecules are correct.

Table 23. Evaporation percentages (%OFF) of the simulated distillation generated by the representative molecules and the experimental sample of GOV and BAP.

SARA	%Off	P.E. T (°C)	P.E.E (°C)	Difference <=5	Diff (%)
SM1	29	404,3	404,3	0	0,0
SM2	11	358,9	358,7	-0,2	0,1
ARM1	35	412,9	413,1	0,2	0,0
ARM2	56	437,6	437,5	-0,1	0,0
ARM3	32	408,6	408,7	0,1	0,0
ARM4	72	456,3	456	-0,3	0,1
ARM5	42	421,9	421,7	-0,2	0,0
RM1	89	481,6	481,6	0	0,0
RM2	94	493,7	493,8	0,1	0,0
RM3	74	458,5	458,6	0,1	0,0
SM1	15	473,7	473,4	-0,3	0,1
SM2	39	566,3	565,3	-1	0,2
ARM1	14	469,9	470,8	0,9	0,2
ARM2	18	481,6	482,9	1,3	0,3
ARM3	34	539,6	539,8	0,2	0,0
ARM4	27	509,2	510	0,8	0,2
ARM5	49	610,9	611,1	0,2	0,0
RM1	30	521,4	521,3	-0,1	0,0
ASM1	61	661,1	662,7	1,6	0,2
ASM2	62	668,1	667,8	-0,3	0,0
ASM3	58	650,4	649,8	-0,6	0,1
ASM4	45	593,7	593,3	-0,4	0,1

Figura 3. Average molecular structure of GOV and BAP

For the case of vacuum gas oil, the molecular structures assumed for the saturated fraction are of the type shown in Figure 10.

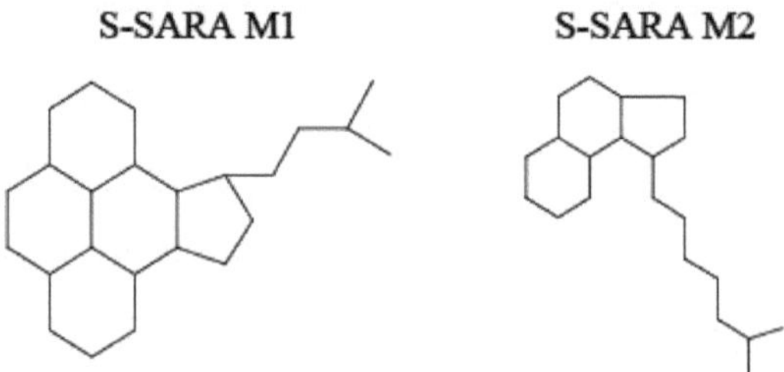

Figura 10. Model molecule representative of GOV saturates.

The S-SARA M1 molecule has a molecular formula of $C_{24}H_{40}$ with a core of 4 cycloparaffins of 6 carbon atoms, plus a 5-membered cycloparaffin linked to which has a short alkyl chain. The S-SARA M2 molecule consists of $C_{21}H_{38}$, with a core possessing 2 6-membered cycloparaffins and a 5-membered one linked to an R of 8 carbon atoms. Short and medium chains were assumed, as this vacuum gas oil comes from a severe thermal cracking process. In the case of aromatics the proposed molecules will be of the type shown in Figure 11. These structures were proposed based on the fact that they are the most common structures found in GOV according to references [lxiii, lx].

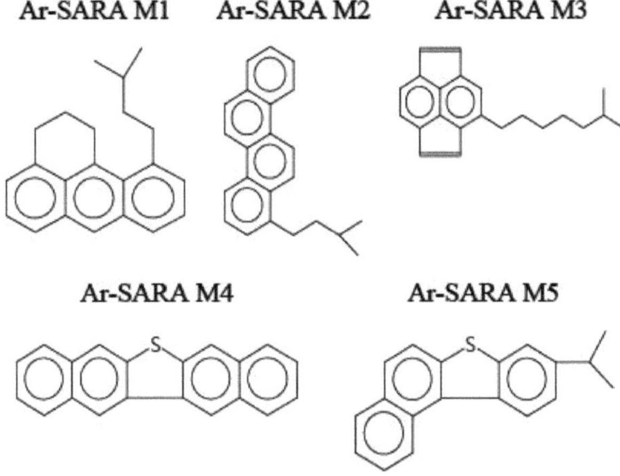

Figura 11. Model molecules representative of the fraction of aromatics for GOV.

The aromatic molecule Ar-SARA M1 $C_{22}H_{24}$, consists of a triaromatic nucleus, which has an alkyl chain of 5 carbon atoms and a saturated chain, likewise, the Ar-SARA M2 molecule of formula $C_{23}H_{22}$ has an alkyl system of 5 carbon atoms and an aromatic nucleus of 4 rings of the peri-condensed type. On the other hand, the Ar-SARA M3 structure whose molecular formula is $C_{22}H_{24}$, consists of 5- and 6-membered aromatic rings in cata-condensed form with an alkyl chain of 8 carbon atoms, the Ar-SARA M4 and Ar-SARA M5 molecules of half-developed chemical formulae $C_{20}H_{12}S$ and $C_{19}H_{16}S$, respectively, consist of dibenzothiophene-type structures.

Figure 12 shows the proposed structures representing the resins contained in the GOV:

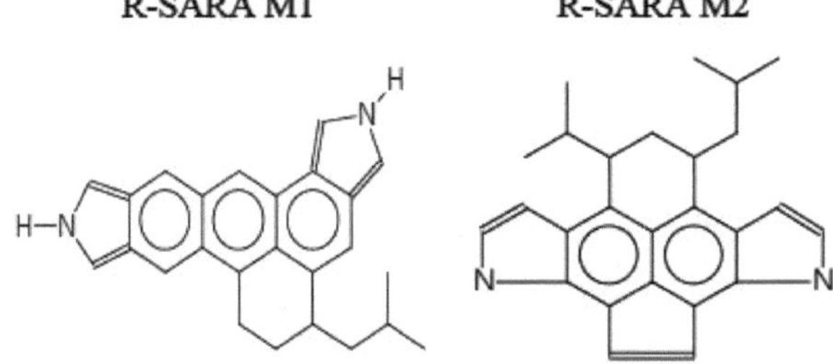

R-SARA M3

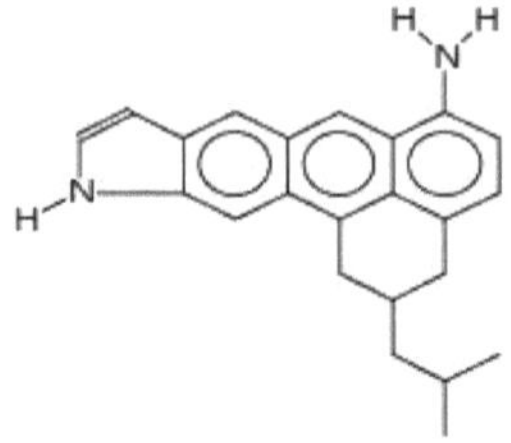

H H
N
H N

Figura 12. Model molecules representative of the resin fraction.

In the case of the resin group molecules, R-SARA M1; R-SARA M2 and R-SARA M3, whose molecular formulae are C25H24N2, C26H28N2 and C23H24N2, it was established that they have mainly nuclei of 2 to 3 aromatic rings of the six-membered peri-condensed type. These nuclei in turn are linked by alkyl chains of 3 to 4 carbon atoms and also possess naphthenic rings.

Finally, it was assumed that the resins possess one hundred percent of the nitrogen atoms present. Since the experimental S.A.R.A. analyses did not report the existence of the asphaltene fraction, it was not taken into account for the study of the vacuum gas oil molecules.

On the other hand, for the case of petroleum tar pitch, the molecular structures proposed for the saturated fraction are of the type shown in Figure 13.

The S-SARA M1 molecule of BAP, with a molecular formula of C30H56, has a structure of 3 monocycloparaffinic nuclei with alkyl branches consisting of 1-2 carbon atoms linked by two-membered alkyl chains. The S-SARA M2 molecule consists of C36H66, represented by 2 tricycloparaffinic nuclei with alkyl chains of 1-2 carbon atoms linked by chains of 4 carbon atoms.

This section was represented by short alkyl chains attached to naphthenic rings, which produced loose molecules capable of rearrangement when subjected to thermal cracking conditions.

S-SARA M1

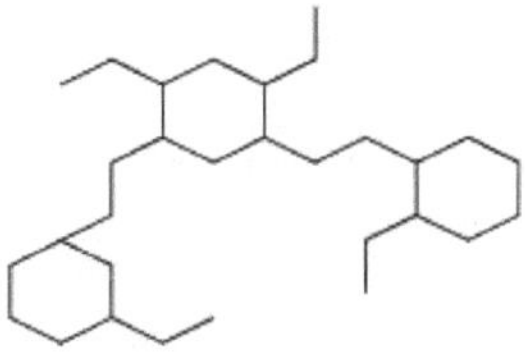

S-SARA M2

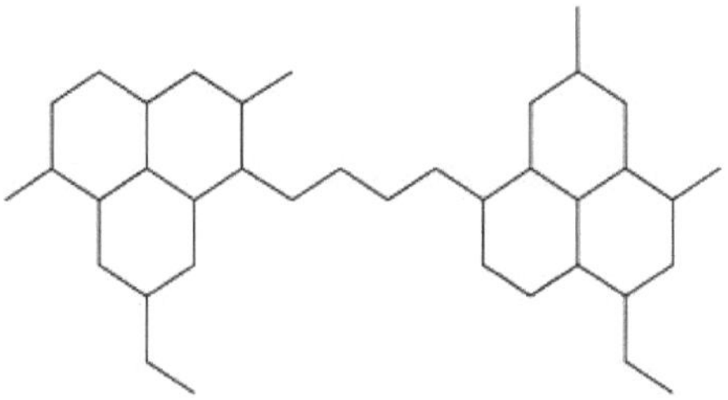

Figura 13. Model molecule representative of the saturated moiety of the GOV.

In the case of BAP aromatics the proposed molecules were of the type shown in Figure 14. As in GOV, the molecules were proposed based on the fact that they are the most common structures found in BAP according to references [lvi, lvii]. The aromatic Ar-SARA M1 molecule consists of a pentaaromatic core whose arrangement is a mixture of cata- and peri-cata-type structures, its semi-developed formula is $C_{23}H_{13}$, the Ar-SARA M2 molecule of $C_{24}H_{14}$ was sketched in a structure of 6 aromatic rings highly concentrated among them. The Ar-SARA M3 structure of formula $C_{28}H_{16}$ consists of 5- and 6-membered aromatic rings without branching.

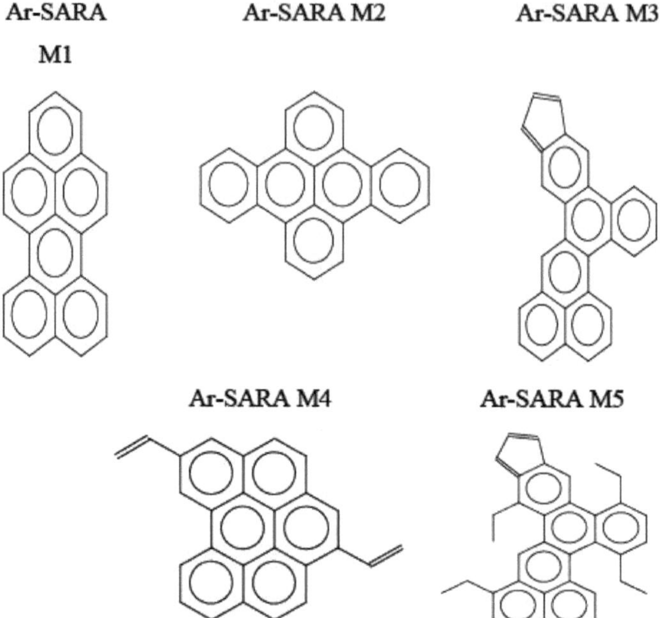

Figura 14. Model molecules representative of the aromatics part for GOV.

As for the Ar-SARA M4 molecule $C_{26}H_{16}$, cata-condensed ordering was observed on a 6-membered aromatic core with olefinic chain branching. Lastly, the Ar-SARA M5 structure of $C_{36}H_{32}$ consists of 5- and 6-membered sleeves with branching of 2 carbon atoms. As for the R-SARA M1 $C_{26}H_{15}N$ resins, it was established that they have a pyridinic system surrounded on one side by a cata-condensed treta-aromatic core and on the other side by a diaromatic system (Figure 15).

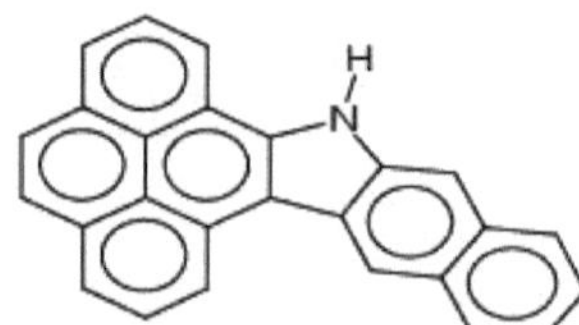

Figura 15. Model molecules representative of the resin part.

In the case of asphaltenes for BAP (Figure 16):

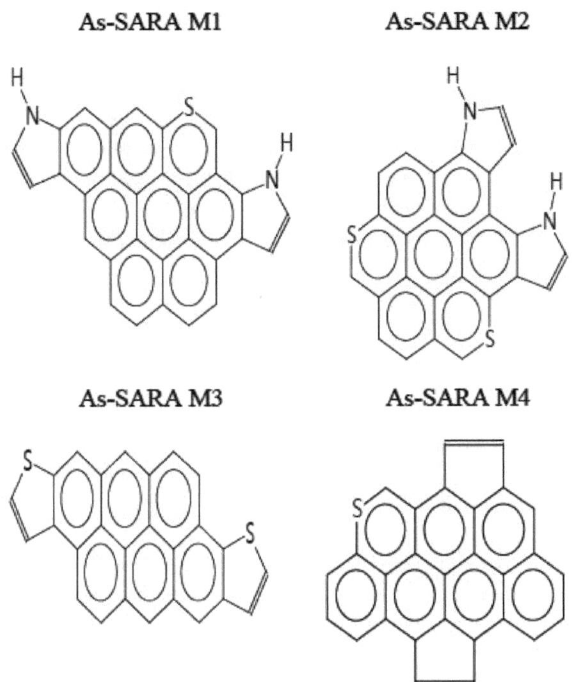

Figure 16. Model molecules representative of the asphaltene fraction.

It was established that they possess nuclei of 6 or more cata-condensed aromatic rings, furthermore it was assumed that the asphaltenes possess 98 % of the total heteroatoms of the charge, their molecular formulae are: C30H14N2S, C26H12N2S2, C26H12S2 and C28H12S.

3.5.2. Balance of the General Equation of BAP Formation

This section presents the general balanced equation for the formation of petroleum tar pitch (PTP) using gas oil under vacuum (GOV) as the starting reagent in the presence of heat and nitrogen (N2) pressure:

$$\begin{aligned}&16C_{24}H_{40}+2C_{21}H_{38}+2C_{23}H_{22}+4C_{22}H_{24}+\\&24C_{20}H_{12}S+8C_{19}H_{16}S+8C_{25}H_{24}N_2+8C_{26}H_{28}N_2\\&+8C_{23}H_{24}N_2\xrightarrow{\Delta}2C_{30}H_{56}+2C_{38}H_{66}+\\&2C_{23}H_{13}+2C_{24}H_{14}+2C_{28}H_{16}+2C_{36}H_{32}\\&+32C_{26}H_{15}N+4C_{30}H_{14}N_2S+4C_{26}H_{12}N_2S_2+\\&4C_{26}H_{12}S_2+4C_{28}H_{12}S+102CH_4+175H_2+8H_2S\end{aligned}\quad(32)$$

The simplified single molecule general equation for GOV and BAP, obtained by summing the representative molecules of vacuum gas oil and petroleum tar pitch respectively, is presented below to simplify the thermodynamic model so that less data is introduced into the model.

$$C_{1784}H_{1880}N_{48}S_{32}\xrightarrow{\Delta}C_{1682}H_{1106}N_{48}S_{24}+102CH_4+175H_2+8H_2S\quad(33)$$

From this equation it can be inferred that in general terms the pitch molecule has a much higher carbon-hydrogen ratio than vacuum gas oil, in addition, a release of gases attributed to thermal conversion could be observed, which contrasts with what was observed experimentally.

3.5.3. Approach to Thermodynamic Modelling

The thermodynamic model approach was based on the assumption that during the thermal conversion process all the material representing the charge or reagent under ideal conditions is converted to form petroleum tar pitch and a fraction of gases, so that the established yield was around 1OO per cent. Based on these assumptions, the following equation was proposed:

$$GOV \xrightarrow{\Delta} BAP + Gases \qquad (34)$$

Where:
GOV= Reactive.
BAP= Main product.
Gases= By-product.

The gases resulting from the breakdown of the charge molecules by the action of temperature were assumed to be methane (CH_4), molecular hydrogen (H_2) and hydrogen sulphide (H_2S), since a loss of mass attributed to the production of these gases was observed in these processes.

In order to study the thermodynamic model representing the conversion of GOV to oil tar pitch, it is necessary to define the heat or enthalpy of reaction change associated with the system under study. From equation 34 it could be said that the change in the enthalpy of reaction ΔH_R is:

$$\Delta H_R = \Delta H_f(BAP_{(s)}) - \Delta H_f(GOV_{(l)}) \qquad (35)$$

Where:

$\Delta H_f(BAP_{(s)}) =$ Variation of the enthalpy of formation of oil tar pitch in solid state.

$\Delta H_f(GOV_{(l)}) =$ Variation of the enthalpy of formation of vacuum gas oil in liquid state.

Then you have to:

$$\Delta H_f(GOV_{(l)}) = \Delta H_f(GOV_{(g)}) - \Delta H_{vap}(GOV) \qquad (36)$$

$$\Delta H_f(BAP_{(s)}) = \Delta H_f(BAP_{(g)}) - \Delta H_{vap}(BAP) - \Delta H_{fus}(BAP) \qquad (37)$$

Where:

$\Delta H_f(GOV_{(g)}) =$ Variation of the enthalpy of formation of vacuum gas oil in gaseous state gaseous state.

$\Delta H_f(GOV) =$ Variation of the enthalpy of vaporisation of vacuum gas oil.

$\Delta H_f(BAP_{(S)}) =$ Variation of the enthalpy of formation of oil tar pitch in solid state.

$\Delta H_{vap}(BAP) =$ Variation of the enthalpy of vaporisation of coal tar pitch oil.

$\Delta H_{fus}(BAP) =$ Variation of the enthalpy of fusion of petroleum tar pitch.

Knowing that:

$$dS = \frac{dQ}{T} \quad (38)$$

Where:

$dS =$ Entropy differential.

$dQ =$ Heat differential.

$T =$ System temperature.

It can be assumed that ΔH_{vap} y ΔH_{fus} in relation of T:

$$\Delta S_{vap} = \frac{\Delta H_{vap}}{T} \quad (39)$$

$$\Delta S_{fus} = \frac{\Delta H_{fus}}{T} \quad (40)$$

Where:

$\Delta S_{VAP} =$ Variation of the entropy of vaporisation.

$\Delta S_{fus} =$ Variation of the entropy of fusion.

But according to Trouton's rule [lxiv]:

$$\Delta H_{vap}(GOV) \cong 21\bar{T}(GOV) \quad (41)$$

$$\Delta H_{vap}(BAP) \cong 21\bar{T}(BAP) \quad (42)$$

$$\Delta H_{vap}(BAP) \cong Cp(BAP) * (T_{(ablandamiento)} - T_o) \quad (43)$$

Where:

$\Delta H_{vap}(GOV) -$ Variation of the enthalpy of vaporisation of vacuum gas oil.

$Cp(BAP) =$ Specific value at constant pressure of petroleum tar pitch.

On the other hand, for the calculation of the entropy variation of ΔS_R:

$$\Delta S_R = S_f(BAP_{(S)}) - S_f(GOV_{(l)}) \quad (44)$$

In which:

$S_f(BAP_{(S)}) =$ Entropy of formation of petroleum tar pitch in solid state.

$S_f(GOV_{(l)}) =$ Entropy of formation of vacuum gas oil in liquid state.

$$S_f(BAP_{(s)}) = S_f(BAP_{(g)}) - \Delta S_{vap}(BAP) - \Delta S_{fus}(BAP) \quad (45)$$

$$S_f(GOV_{(l)}) = S_f(GOV_{(g)}) - \Delta S_{vap}(GOV_{(l)}) \quad (46)$$

Where:

$S_f(BAP_{(g)}) =$ Entropy of formation of petroleum tar pitch in the gaseous state.

$\Delta S_{vap}(BAP) =$ Vaporisation entropy variation of petroleum tar pitch.

$\Delta S_{fus}(BAP) =$ Variation of the entropy of fusion of petroleum tar pitch.

$S_f(GOV_{(g)}) =$ Entropy of formation of vacuum gas oil in gaseous state.

$\Delta S_{vap}(GOV_{(l)}) =$ Vaporisation entropy variation of vacuum gas oil in liquid state.

Assuming that the variation of the enthalpy of vaporisation ΔH_{vap} as a ratio of T as:

$$\Delta S_{vap}(GOV) = \frac{\Delta H_{vap}(GOV)}{\bar{T}_{ebullición}} \quad (47)$$

Where:

$$\bar{T}_{ebullición} = \sum_i \frac{\alpha i}{\alpha T} * Teb_i^{GOV} \quad (48)$$

Likewise, for the calculation of $\Delta S_{vap}(BAP)$, the following relationship is taken:

$$\Delta S_{vap}(BAP) = \frac{\Delta H_{vap}(BAP)}{\bar{T}_{ebullición}} \quad (49)$$

Where:

$$\bar{T}_{ebullición} = \sum_i \frac{\beta i}{\beta T} * Teb_i^{BAP} \quad (50)$$

For the calculation of $\Delta S_{vap}(BAP)$, it follows that:

$$\Delta S_{fus}(BAP) = \frac{\Delta H_{fus}(BAP)}{T(ablandamiento)} \quad (51)$$

In which:

$$\Delta Hfus \cong Cp(T_{(ablandamiento)} - T_0) \quad (52)$$

On the other hand, for the calculation of the variation of the Gibbs free energy of the reaction ΔG_R: it was assumed that:

$$\Delta G_R = \Delta H_R - T_{(proceso)} * \Delta S_R \quad (53)$$

$$K_{equilibrio} = e^{\frac{-\Delta G_R}{RT}} \quad (54)$$

Where:

$\Delta S_R =$ Variation of reaction entropy.

$\Delta H_R =$ Variation of the enthalpy of reaction.

$K_{equilibrio} =$ Equilibrium constant of the reaction.

3.5.4. Calculation of Cp (a, b, c and d), Enthalpy of Formation and Entropy of Formation for each molecule of GOV and BAP at Ideal Gas State by QBTherm™ V3.0 software.

Table 24 shows the results of the calculations of the polynomial coefficients for the specific heat at constant pressure Cp (a, b, c and d), Enthalpy of Formation and Entropy of Formation for each of the molecules proposed for GOV at ideal gas state by the QBTherm™ V3.0 program. [lxv, lxvi].

Table 24. Enthalpy of formation, entropy of formation and Cp (a, b, c and d) for each GOV molecule.

for each GOV molecule.

Molecule	ΔHí (Kcal)	Sf (Cal)	A	b	C	d
S-M1	-88,70	83,50	96,20	0,27	-8,00E-05	-6,26E+06
S-M2	-87,74	164,08	55,65	0,29	-9,00E-05	-4,53E+06
Ar-M1	17,31	178,29	60,17	0,20	-5,00E-05	-3,36E+06
Ar-M2	43,23	172,91	66,35	0,19	-5,00E-05	-3,50E+06
Ar-M3	17,05	153,80	67,85	0,19	-5,00E-05	-3,27E+06
Ar-M4	90,40	160.14	49.56	0.16	-6.00E-05	-2.57E+06
Ar-M5	53,10	161.57	49.65	0.17	-5.00E-05	-2.55E+06
R-M1	61,44	146,67	80,48	0,22	-6,00E-05	-4,47E+06
R-M2	33,86	138.71	90.04	0.23	-6.00E-05	-5.00E+06
R-M3	43,19	151.64	64.33	0.23	-7.00E-05	-3.58E+06

Table 25 presents the values of Cp, Enthalpy of Formation and Entropy of Formation corrected by the stoichiometric coefficient in order to take these data to establish the physicochemical parameters of a general vacuum gas oil molecule.

Table 25. Enthalpy of formation, entropy of formation and Cp (a, b, c and d) for each GOV molecule corrected by stoichiometric coefficient.

each GOV molecule corrected by the stoichiometric coefficient.

Molécu the	C.E	ΔHí (Kcal)	Sf (Cal)	a	B	c	d
S-M1	16	-1419,20	1336,07	1539,18	4,38498	-1,31E-03	-1,00E+08
S-M2	2	-175,48	328,16	111,29	0,58440	-1,80E-04	-9,06E+06
Ar-M1	2	34,62	356,59	120,34	0,40233	-1,00E-04	-6,73E+06
Ar-M2	2	86,46	345,82	132,70	0,38202	-1,10E-04	-7,00E+06
Ar-M3	2	34,10	307,60	135,70	0,38668	-1,10E-04	-6,54E+06
Ar-M4	24	2169,60	3843,32	1189,34	3,83987	-1,32E-03	-6,16E+07
Ar-M5	8	424,80	1292,59	397,17	1,37052	-4,60E-04	-3,58E+07
R-M1	8	491,52	1173,33	643,82	1,73819	-5,00E-04	-4,00E+07
R-M2	8	270,88	1109,67	720,31	1,87290	-5,20E-04	-2,87E+07
R-M3	8	345,52	1213,14	514,66	1,85579	-5,20E-04	-4,82E+07

Table 26 shows the results of the calculations of Cp (a, b, c and d), Enthalpy of Formation and Entropy of Formation for each of the molecules proposed for BAP in the ideal gas state by the QBTherm ™ V3.0 program.

Table 26. Enthalpy of formation, entropy of formation and Cp (a, b, c and d) for each BAP molecule.

for each BAP molecule.

Molecule	AHÍ (Kcal)	Sf (Cal)	a	b		d
S-M1	-149.44	223.45	86.87	0.42	-0.00013	-6030475
S-M2	-170,72	198,18	92,78	0,55	-0,00017	-9087887
Ar-M1	100,07	121,29	61,96	0,16	-0,00005	-3298526
Ar-M2	101,70	148,50	64,47	0,16	-0,00005	-3462912
Ar-M3	119,00	169,15	75,35	0,19	-0,00006	-4024130
Ar-M4	146,02	216,94	58,72	0,14	-0,00004	-3098846
Ar-M5	69,24	243,50	97,66	0,30	-0,00007	-4768258
R-M1	121,04	99,38	74,71	0,18	-0,00006	-3928510
As-M1	166,84	130,10	87,71	0,22	-0,00007	-4198376
As-M2	157,18	141,70	72,81	0,21	-0,00007	-3301746
As-M3	135,88	183,45	63,19	0,21	-0,00008	-2930099
As-M4	141,34	148,90	73,05	0,20	-0,00007	-3497038

Table 27 presents the values of Cp (a, b, c and d), Enthalpy of Formation and Entropy of Formation corrected by the stoichiometric coefficient for the proposed petroleum tar pitch molecules. These calculations will be used to evaluate the average properties of the overall BAP molecule.

Table 27. Enthalpy of formation, entropy of formation and Cp (a, b, c and d) for each BAP molecule corrected by stoichiometric coefficient.

each BAP molecule corrected by the stoichiometric coefficient.

Molécula	C.E	ΔHí (Kcal)	Sf (Cal)	a	b	c	d
S-M1	2	-298,88	446,89	173,75	0,83	-0,00025	-12060950
S-M2	2	-341,44	396,3 6	185,56	1,11	-0,00035	-18175774
Ar-M1	2	200,14	242,59	123,91	0,31	-0,00010	-6597052
Ar-M2	2	203,40	297,01	128,93	0,33	-0,00010	-6925824
Ar-M3	2	238,00	338,31	150,71	0,38	-0,00012	-8048260
Ar-M4	2	292,04	433,87	117,45	0,28	-0,00007	-6197692
Ar-M5	2	138,48	486,99	195,31	0,60	-0,00015	-9536516
R-M1	32	3873,28	3180,16	2390,68	5,78	-0,00178	125712320
As-M1	4	667,36	520,39	350,83	0,88	-0,00030	-16793504
As-M2	4	628,72	566,80	291,24	0,86	-0,00030	-13206984
As-M3	4	543,52	733,81	252,76	0,84	-0,00031	-11720396
As-M4	4	565.36	595.62	292,20	0,79	-0.000	-13988

Table 28 shows the data obtained by means of the group addition method which was taken from the bibliographic source Perry's HandBook.

Table 28. Enthalpy of formation, entropy of formation and Cp (a, b, c and d) for each gas molecule.

for each gas molecule.

Molecule	ΔHf (Kcal)	Sf (Cal)	a	b	c	d
CH_4	-17,90	44,49	2,86	0,02	-0,000005	30193
H_2	0	31,21	4,04	0,01	0	0
H_2S	-4,93	49,16	8,18	0	0	0

3.5.5. Calculation of Cp (a, b, c and d), enthalpy of formation and entropy of formation for the general GOV and BAP molecule at ideal gas state

From the results of the physicochemical properties corrected by the stoichiometric coefficient in the previous section, a general result of Cp (a, b, c and d), enthalpy of formation and entropy of formation at ideal gas state for vacuum gas oil and petroleum tar pitch could be obtained in order to simplify the reasoning. These are denoted in Table 29 below. As in the GOV, the molecules were proposed based on the fact that they are the most common structures found in BAP according to references [lvi, lvii].

Table 29. Enthalpy of formation, entropy of formation and Cp for general GOV and BAP molecules.

for general GOV and BAP molecules.

Molecule	ΔHf (Kcal)	Sf (Cal)	a	B	C	d
GOV	28,29	141,33	68,81	1,05	0	-42964449
BAP	108,23	132,88	75,05	0,21	0	-4015539
CH_4	-17,90	44,49	2,86	0,02	-0,000005	30193
H_2	0	31,21	4,04	0,01	0	0
H_2S	-4,93	49,16	8,18	0	0	0

The aromatic molecule Ar-SARA M1 consists of a pentaaromatic nucleus whose arrangement is a mixture of cata- and peri-cata-type structures, its semi-developed formula is $C_{23}H_{13}$, the Ar-SARA M2 molecule of $C_{24}H_{14}$ was outlined in a structure of 6 aromatic rings highly concentrated among themselves. The Ar-SARA M3 structure of formula $C_{28}H_{16}$ consists of 5- and 6-membered aromatic rings without branching.

As shown in Table 29, the net difference in the heat of formation between reactant and product is very positive, so the driving force of BAP formation from GOV is attributed to the fact that large amounts of short hydrocarbon moles are formed, such as CH4 whose free energy of formation is very negative, thus favouring the pitch formation process.

3.5.6. Calculation of Enthalpy, Entropy and Free Energy of Reaction as a Function of Pressure and Temperature for the Simplified General Chemical Equation of Formation of BAP

Table 30 presents the results of Enthalpy of Formation, Entropy of Formation and Free Energy of Reaction as a function of pressure and temperature for the simplified general chemical equation representing the formation of petroleum tar pitch. These results were obtained through the QBTherm™ V3.0 program.

When comparing the temperature conditions used in the development of the experimental work it can be said that the higher the temperature, the more thermodynamically stable the system will be, which can be corroborated by the variation of the ΔG of reaction.

Table 30. Enthalpy of formation, entropy of formation and free energy of reaction as a function of pressure and temperature.

as a function of pressure and temperature.

TEMP (°C)	Pressure (psig)	ΔH (Kcal)	ΔS (Cal)	ΔG (Kcal)
420	250	-691.1	11034.0	-7219.7
430	250	-657.3	11082.5	-7314.1
440	250	-623.1	11130.8	-7409.1

In general terms, the process of petroleum tar pitch production can be said to be irreversible and spontaneous. At any temperature the chemical reaction is exothermic favouring the formation of products. Knowing that in Table 29 the difference in entropy of formation between the reactant and the products is small and analysing Table 30 we can say that the calculation between the entropy change of the reaction is large and very positive attributed to the large amount of gases generated during thermal cracking. In other words, in a process where one solid produces another solid plus gases, the entropy change is large, because gases have more degrees of freedom than solids.

One might think that this type of process would work more efficiently at lower pressures with the idea of shifting via Le Chatelier the equilibria to the products. However, an excess in the production of gases would increase the generation of products towards pitch, but also towards coke. Therefore, from a thermodynamic point of view, the coke formation process will be favoured over any other process, so for its control, the system should be pressurised, favouring the production of pitch. When analysing the experimental results according to the thermodynamic model proposed for the conversion of GOV into BAP, it is established that, in general, the theoretical process effectively represents the experimental conversion conditions. Furthermore, the proposed molecules of both the charge and the product are below the 5 % error, which guarantees its effectiveness.

A small difference between both studies is that it was assumed that the reagent is transformed directly to pitch without forming pitch base, because this is a transient stage in which only a distillation process occurs, where a fraction of light is produced, this was done in order to facilitate the thermodynamic calculations, however this difference between both studies, does not contribute significant changes to the overall balance of the conversion, since it is only a physical separation which does not modify the chemical structure of the molecules within the thermal conversion.

The main limitation of the proposed thermodynamic model is that it does not take the chemical stability of the free radicals formed in the system as a variable. Likewise, as it is only of a thermodynamic nature, the reaction time is not taken into consideration, which as it was experimentally evidenced is an essential factor at the time of the formation of the expected product and because this model only allows calculating the physicochemical variables of state (ΔH, ΔS and ΔG) the maximum conversion of the system at different conditions of temperature and pressure will be reached by using the following chemical equation to calculate the equilibrium constant of the system. Likewise, when calculating the variation of the Gibbs free energy for a given condition, this calculated value was very high, which means that the equilibrium constant is higher, implying that the conversion of the product increases as the temperature increases, but is limited by the formation of coke.

When comparing the results observed for the 420 °C condition, by means of the thermodynamic model, it is observed that there is a high Gibbs free energy, although experimentally there was no appreciable change, this can be attributed to the fact that the activation energy for the bond breaking was not enough, so the molecules of the charge did not pass to the gas phase where the breaking and subsequent structural rearrangement of the molecules occurs. Therefore, it should be noted that the longer the reaction time, the greater the release of free radicals. On the other hand, we can say that thermal cracking reactions are thermodynamically favoured at temperature conditions above 300 °C but they are limited by the reaction time.

CHAPTER 4

CONCLUSIONS ON PRODUCTION DE BREA

From the results obtained from the experiments carried out, it can be concluded that:

- The increasing trend of reactant properties towards product demonstrated the technical feasibility of producing highly condensed aromatic products such as petroleum tar pitch from GOV.
- The range of variables selected have an effect on both the product yield and the overall properties of the petroleum tar pitch, showing that the reaction temperature and reaction time directly affect the physicochemical properties essential for the production of pitch from vacuum gas oil.
- The optimum operating conditions to obtain the highest yield of oil tar pitch with the best physico-chemical properties (softening point and micro-coal residue) are reaction temperature 440 °C, reaction time 45 to 60 minutes and reaction pressure 250 psig.
- At moderate temperature and residence time conditions (430-440 °C and 45-60 min), high thermal cracking reactions occur, which break down the starting hydrocarbon molecules generating free radicals that form higher molecular weight fractions by polymerisation, condensation, dehydrogenation and dealkylation, in turn producing light fractions.
- The presence or not of chemical reactions in the system under study can be determined by the presence of two phases after the reaction: gaseous and solid.
- The increasing trend in the percentage of toluene insolubles and microchar residue is directly proportional to the residence time and reaction temperature.
- The higher the distillation temperature, the higher the concentration of polyaromatic species, whose carbon-hydrogen ratio tends to be higher than that of the starting product.
- Due to the thermal effect, there is a variation in the percentages of saturates, asphaltenes, resins and aromatics, which is consistent with the high variation of physical parameters such as density and viscosity of both the filler and the product.
- The increase in boiling points of the product relative to the reactant is directly proportional to the thermal conversion of the charge, since the higher the boiling point, the greater the complexity of the associated molecules.
- The effectiveness of the mass balance performed for representative molecules designed for GOV and BAP is verified, since the difference between the percentage of the theoretical analysis and the experimental value does not exceed 2 percent error, which is within the expected range (less than or equal to 5 %).
- Vacuum gas oil molecules are highly aromatic structures of about 80%, with a low carbon-hydrogen ratio, composed of polyunsaturated and polyaromatic nuclei with no more than four structural rings and alkyl chains. Their SARA fraction does not have the asphaltene fraction.
- In general terms, the pitch molecule has a much higher carbon/hydrogen (C/H) ratio than vacuum gas oil, composed of polyunsaturated and polyaromatic nuclei with between three and eight structural rings of six carbon atoms. It has a low percentage of resins.
- The net difference of the heat of formation between reactant and product is very positive for so the driving force of BAP formation from GOV is attributed to the fact that large amounts of short hydrocarbon moles are formed, such as CH4 whose free energy of formation is very negative favouring the pitch formation process.

- The variation of the Gibbs free energy for the reaction corroborates that the higher the temperature, the higher the thermodynamic stability of the system.
- The process of petroleum tar pitch production is irreversible and spontaneous. At any temperature the chemical reaction is exothermic favouring the formation of products.
- Excess gas production increases product generation towards pitch, but also towards coke. This is why from a thermodynamic point of view the coke formation process will be favoured over any other process, so for its control the system should be pressurised to an inert atmosphere (N_2) favouring the production of pitch.
- During thermal cracking the entropy change of the reaction is large and very positive attributed to the large amount of gases generated due to the fact that gases have more degrees of freedom than solids. When analysing the experimental results in terms of the thermodynamic model proposed for the conversion of GOV into BAP, it is established that in general line
- The theoretical process effectively represents the experimental conversion conditions, the difference between the two studies does not bring significant changes to the overall conversion balance.
- The main limitation of the proposed thermodynamic model is that it does not take into account the chemical stability of the free radicals formed in the system. Also, as it is only of a thermodynamic nature, the reaction time is not taken into consideration,
- The calculated Gibbs free energy change for the system is high, which results in a higher equilibrium constant, implying that the product conversion increases with increasing temperature, but is limited by the formation of coke.
- The bond-breaking tendency implies that the longer the reaction time, the greater the release of free radicals, which translates into a greater formation of chemical structures different from the initial hydrocarbons.

ABOUT THE AUTHOR

Dr. Jairo José Rondón Contreras. Associate Professor of Engineering at the Polytechnic University of Puerto Rico. Degree in Chemistry, Master and PhD in Applied Chemistry, mention in Materials Studies from the Universidad de Los Andes. Since 2007 he has worked in the oil and gas industry, acquiring experience in the area of applied chemical processes; he has also developed basic and detailed engineering projects for the chemical and petroleum industry. His research interests include Materials Science and Engineering, Engineering Design, Applied Chemistry, and Project Management. Author and co-author of scientific articles on chemical and biomedical engineering, solids characterisation, kinetics, catalysis, and thermodynamics.

BIBLIOGRAPHY

[i] Speight, J. G. (2014). The chemistry and technology of petroleum. CRC press.

[ii] Rondón, J. (2011). Hydrodesulphurisation of dibenzothiophene on Mo catalysts supported on the nanoporous material MCM-48. Special Degree Work (MSc). University of Los Andes, Faculty of Sciences. Mérida- Venezuela.

[iii] Barberii, E. E. (1998). The illustrated well. PDVSA, Programa de Educación Petrolera. Venezuela.

[iv] Rondón, J., Meléndez, H., Lugo, C., del Castillo, H., & Imbert, F. (2016). Synthesis, characterisation and catalytic activity of MoS2/MCM-48 in the hydrodesulphurisation of dibenzothiophene. Advances in Chemistry, 11(1), 35-45.

[v] Rondón, J., Lugo, C., Meléndez, H., Pérez, P., Del Castillo, H., & Imbert, F. (2021). Study of NiMoS2/MCM-48 type heterogeneous solids and catalytic activity in the hydrodesulphurisation of dibenzothiophene. Science and Engineering Journal. Vol, 42(2).

[vi] Department of Chemistry (2001). Organic Chemistry Laboratory Manual. Universidad de Los Andes, Mérida, Venezuela.

[vii]Furniss, B. S. (1989). Vogel's textbook of practical organic chemistry. Pearson Education India.

[viii] Oropeza, F. (2004). Obtaining Flekes from the Bottom Oil of the Hot Separator of the HDH PLUS Hydroconversion Process® (Internship Report). Universidad Central de Venezuela, Caracas, Venezuela.

[ix] Hume, S. M. (1993). Influence of raw material properties on the reactivity of carbon anodes used in the electrolytic production of aluminium. AluminiumVerlag.

[x] Durán, Y. (2006). Vapoconversion reactions catalysed by nanostructured systems based on transition and alkali metals (Undergraduate thesis). PDVSA-INTEVEP. S.A., Universidad de Los Andes, Mérida, Venezuela.

[xi] Jones, D. S., Pujadó, P. P. (Eds.) (2006). Handbook of petroleum processing. Springer Science & Business Media.

[xii]Gray, R. M. (1994). Upgrading petroleum residues and heavy oils. CRC press.

[xiii] Thiel, P. A., Madey, T. E. (1987). The interaction of water with solid surfaces: Fundamental aspects. Surface Science Reports, 7(6-8), 211-385.

[xiv] Rahimi, P. M., Gentzis, T. (2006). The chemistry of bitumen and heavy oil processing. In Practical Advances in Petroleum Processing (pp. 597-634). Springer New York.

[xv]Kossiakoff, A., Rice, F. O. (1943). Thermal decomposition of hydrocarbons, resonance stabilization and isomerization of free radicals. Journal of the American Chemical Society, 65(4), 590-595.

[xv]Schabron, J. F., Pauli, A. T., Rovani, J. F. (2002). Residua coke formation predictability maps. Fuel, 81(17), 2227-2240.

[xvii] Li, S., Liu, C., Liang, W. (2001). Coke formation in thermal and catalytic cracking of Shengli and Gudao vacuum residue in view of colloidal stability. Preprints-American Chemical Society. Division of Petroleum Chemistry, 46(3), 259-263.

[xviii] León, O. G., Sosa, Y. R., Álvarez, C. M. (2017). Industrial applications. Revista de Ingeniería, 38(2), 115-123.

[xix] Rice, F. O., Stallbaumer, A. L. (1942). The Decomposition of Cyclohexene Oxide and

1, 4-Cyclohexadiene from the Stand-point of the Principle of Least Motion1. Journal of the American Chemical Society, 64(7), 1527-1530.

[xx]Rice, F. O.; Roberts, R. (1943). The Structure of Diketene. Journal of the American Chemical Society, 65(9), 1677-1681.

[xxi] Higuerey, I. (2001). Estudio Comparativo de la Distribución de Productos de las Reacciones de Craqueo Térmico y Vapocraqueo Termocatalítico del Residuo Tía Juana Pesado (Doctoral thesis). Universidad Central de Venezuela, Caracas, Venezuela.

[xxii] Clark, P. D., Hyne, J. B., Tyrer, J. D. (1984). Some chemistry of organosulphur compound types occurring in heavy oil sands: 2. Influence of pH on the high temperature hydrolysis of tetrahydrothiophene and thiophene. Fuel, 63(1), 125-128.

[xxiii] Fan, H. F., Liu, Y. J., Zhao, X. F. (2001). Study on composition changes of heavy oils under steam treatment. Journal of Fuel Chemistry and Technology, 29(3), 269-272.

[xxiv] González-Cortés, S., & Imbert, F. E. (Eds.) (2018). Advanced Solid Catalysts for Renewable Energy Production. IGI Global.

[xxv] Aquino, L. (2006). Waste upgrading. Refining course section 9. PDVSA- INTEVEP. Miranda, Venezuela.

[xxvi] Barneto, A. G., Carmona, J. A., Garrido, M. J. F. (2016). Thermogravimetric assessment of thermal degradation in asphaltenes. Thermochimica Acta, 627, 1-8.

[xxvii] Kataria, K. L., Kulkarni, R. P., Pandit, A. B., Joshi, J. B., Kumar, M. (2004). Kinetic studies of low severity visbreaking. Industrial & engineering chemistry research, 43(6), 1373-1387.

[xxviii]Gray, M. R., McCaffrey, W. C. (2002). Role of chain reactions and olefin formation in cracking, hydroconversion, and coking of petroleum and bitumen fractions. Energy & fuels, 16(3), 756-766.

[xxix] Machin, I., De Jesus, J. C., Zacarías, L., Rivas, G., Delgado, O., Sánchez, R., Sardella, R., Higuerey, I. (2006). AQUACONVERSION Mechanism® (Technical Report). PDVSA-INTEVEP. S. A, INT-11129, Miranda, Venezuela.

[xxx] Mottola, M. (1991). Production and Characterisation of Petroleum Coke and Tars (Master's Thesis). PDVSA-INTEVEP. S. A; Universidad Simón Bolívar, Caracas, Venezuela.

[xxxi] Hammond, D. G., Lampert, L. F., Mart, C. J., Massenzio, S. F., Phillips, G. E., Sellards, D. L., Woerner, A. C. (2003). Review of fluid coking and flexicoking technologies. In AIChE 6th Topical Conf. on Refining Processing, USA, Spring.

[xxxii] Romano, M. K., de Chamorro, M. P. (2003). Preliminary study of acid recycling in the simultaneous desmetallisation and desulphurisation of Venezuelan petroleum cokes via microwaves. Revista Facultad. Ing, 18, 73-78.

[xxxiii]Velasco, L., Mota, C., Rodríguez, D. (1990). U.S. Patent No. 4,961,837. Washington, DC: U.S. Patent and Trademark Office.

[xxxiv]Azuaya, A., De Salazar, C., Palazon, E. (2001). European Patent Register E.P. 1 130 077 A2. Madrid, Spain.

[xxxv] Dubois, J., Agache, C., White, J. L. (1997). The carbonaceous mesophase formed in the pyrolysis of graphitizable organic materials. Materials Characterization, 39(2), 105-137.

[xxxvi]Delgado, O. (2005). Petroleum Mesophase as a Precursor of New Carbonaceous Materials (Technical Report). PDVSA-INTEVEP. S.A, INT- 10649, Miranda, Venezuela.

[xxxvii] Nic, M., Hovorka, L., Jirat, J., Kosata, B., Znamenacek, J. (2005).

IUPAC Compendium of Chemical Terminology-The Gold Book. International Union of Pure and Applied Chemistry.
[xxxviii] García-Cuello, V., Vargas-Delgadillo, D., Murillo-Acevedo, Y., Cantillo-Castrillon, M. Y., Rodríguez-Estupiñán, P., Giraldo, L., Moreno- Piraján, J. C. (2011). Thermodynamics of the Interactions Between Gas-Solidand Solid-Liquid on Carbonaceous Materials. In Thermodynamics-Interaction Studies-Solids, Liquids and Gases. InTech.
[xxxix] Zander M. (2000). Chemistry and Properties of Coal-Tar and Petroleum Pitch, in "Science of Carbon Materials. Marsh, H and Rodríguez Reinoso F. University of Alicante. Spain.
[xl] Parker, J. E., Johnson, C. A., John, P., Smith, G. P., Herod, A. A., Stokes, B. J., Kandiyoti, R. (1993). Identification of large molecular mass material in high temperature coal tars and pitches by laser desorption mass spectroscopy. Fuel, 72(10), 1381-1391.
[xli] Greinke, R. A., O'connor, L. H. (1980). Determination of molecular weight distributions of polymerized petroleum pitch by gel permeation chromatography with quinoline eluent. Analytical Chemistry, 52(12), 1877-1881.
[xlii] Simon, H (1989). Methods of Analysis for Electrode Pitch. Allied Corporation Ironton. Ohio-USA.
[xliii] Marsh, H., Kuo, K. (1989). Kinetics and catalysis of carbon gasification, Introduction to Carbon Science (pp. 107-151). University of Newcastle upon Tyne, Newcastle upon Tyne, NE1 7RU, U.K.
[xliv] French Petroleum Institute, Petroleos de Venezuela, S.A., (2007). MSc in Refining, Engineering and Gas. Module 2: Characterization of Petroleum Products. An International Training Organization. Caracas, Venezuela.
[xlv] Forrest, M., Marsh, H. (1981). Composition of pore-wall material in metallurgical coke: Considerations of strength, gasification and thermal stress. Fuel, 60(5), 418-422.
[xlvi] Lubkowitz, J. Ceballo, C. (1996). Course "Simulated Distillation", COLACRO VI, Caracas, Venezuela.
[xlvii] Ceballo, C., Murgia, E. (1981). Simulated Distillation of Gasolines. Revista Técnica INTEVEP 1 (2), 99 -107. PDVSA- INTEVEP. Miranda, Venezuela.
[xlviii] ASTM D2887-16a, (2016). Standard Test Method for Boiling Range Distribution of Petroleum Fractions by Gas Chromatography, ASTM International, West Conshohocken, PA, USA.
[xlix] ASTM D7169-16. (2016). Standard Test Method for Boiling Point Distribution of Samples with Residues Such as Crude Oils and Atmospheric and Vacuum Residues by High Temperature Gas Chromatography, ASTM International, West Conshohocken, PA, USA.
[l] VAPRO® Vapour Pressure Osmometer. 1995, 1998, 1998, 2000, 2002 Wescor, Inc. Printed in Spain. Wescor, Vapro, Optimol Osmocoll, and Blow Clean are registered trademarks of Wescor, Inc. Retrieved from http://www.wescor.com/translations/Translations/M2468-4-ES.pdf.
[li] Bidlingmeyer, B. A. (1992). Practical HPLC methodology and applications. John Wiley & Sons.
[lii] ASTM D5291-16, (2016). Standard Test Methods for Instrumental Determination of Carbon, Hydrogen, and Nitrogen in Petroleum Products and Lubricants, ASTM

International, West Conshohocken, PA, USA.
[1111] Skoog, D. A., Holler, F. J. N. T. A., Timothy, A. D. A. (2001). Principles of instrumental analysis (No. 543.4/. 5). McGraw-Hill Interamericana de España.
[liv] Fischer, W., Mannweiler, U., Keller, F., Perruchoud, R., Buhler, U. (1995). Anodes for the aluminium industry. Sierre, Switzerland, R&D Carbon Ltd.
[lv] ASTM D2622-16, (2016). Standard Test Method for Sulfur in Petroleum Products by Wavelength Dispersive X-ray Fluorescence Spectrometry, ASTM International, West Conshohocken, PA, USA.
[lvi] Kershaw, J. R., Black, K. J. (1993). Structural characterization of coal-tar and petroleum pitches. Energy & fuels, 7(3), 420-425.
[lvii] Qian, S. A., Zhang, P. Z., Li, B. L. (1985). Structural characterization of pitch feedstocks for coke making: use of 13C coupled 1H NMR spectroscopy. Fuel, 64(8), 1085-1091.
[lviii] López, I. (1996). Course "Introduction to Petroleum Refining". Centro *Internacional de Educación y Desarrollo, PDVSA, II* (23), Caracas -Venezuela.
[lx] Ali, F., Khan, Z. H., Ghaloum, N. (2004). Structural studies of vacuum gas oil distillate fractions of Kuwaiti crude oil by nuclear magnetic resonance. Energy & fuels, 18(6), 1798-1805.
[lxi] Levine, I. N. (2004). Physical Chemistry (Vol. 1 and 2).
[lxii] Lyman, W. J., Reehl, W. F., Rosenblatt, D. H. (1990). Handbook of chemical property estimation methods: environmental behavior of organic compounds.
[lxiii] Peters, K. E., Moldowan, J. M. (1993). The biomarker guide: interpreting molecular fossils in petroleum and ancient sediments.
[lxiv] Castellan Gilbert, W. (1998). Physicochemistry. Editorial Addison Wesley Longman. Mexico.
[lxv] Rondón, J., Meléndez, H., Lugo, C., García, E., Barros, D., & Del Castillo, H. (2014). Construction of a formation matrix for the production of anode-grade petroleum tar pitch by thermal cracking. Science and Engineering, 35(3), 115-123.
[lxvi] Rondón, C. J., Meléndez, Q. H., Lugo, G. C., García, M. E., Belandria, L., Barros, B. D., & Del Castillo, H. (2011). Development of a molecular thermodynamic model of the anode-grade petroleum tar pitch production process by vacuum gas oil thermal cracking. Science and Engineering, 32(3), 129-139.

Printed by Books on Demand GmbH, Norderstedt / Germany